The Sequence Theory Foundation of Cosmos Theory and Gambol Theory

Sequence Equation for Cosmos Theory
The Sequence of Sequences
The Four Sequences of Sequence Theory
Coupling Constants Related to Fermion Masses
Exact Fine Structure Constant Eigenvalue α
QED and Other Interactions in the Eigenvalue Condition
Fundamental Fermion Sequences of Mass Values
Power Sequence of Coupling Constants
Sequence Theory Basis of Gambol Theory
Strong and ElectroWeak Effects on Fermion Masses
Cosmos Spaces as Sets of SU(8) Representations

Stephen Blaha Ph. D.
Blaha Research

MMXXV

ISBN: 978-1-7345834-5-8

Rev. 00/00/01   March 24, 2025

To Margaret

Some Other Books by Stephen Blaha

SuperCivilizations: Civilizations as Superorganisms (McMann-Fisher Publishing, Auburn, NH, 2010)

All the Universe! Faster Than Light Tachyon Quark Starships & Particle Accelerators with the LHC as a Prototype Starship Drive Scientific Edition (Pingree-Hill Publishing, Auburn, NH, 2011).

Unification of God Theory and Unified SuperStandard Model THIRD EDITION (Pingree Hill Publishing, Auburn, NH, 2018).

The Exact QED Calculation of the Fine Structure Constant Implies ALL 4D Universes have the Same Physics/Life Prospects (Pingree Hill Publishing, Auburn, NH, 2019).

Passing Through Nature to Eternity ProtoCosmos, HyperCosmos, Unified SuperStandard Theory (Pingree Hill Publishing, Auburn, NH, 2022).

HyperCosmos Fractionation and Fundamental Reference Frame Based Unification: Particle Inner Space Basis of Parton and Dual Resonance Models (Pingree Hill Publishing, Auburn, NH, 2022).

The Cosmic Panorama: ProtoCosmos, HyperCosmos, Unified SuperStandard Theory (UST) Derivation (Pingree Hill Publishing, Auburn, NH, 2022).

God and and Cosmos Theory (Pingree Hill Publishing, Auburn, NH, 2023).

Newton's Apple is Now The Fermion (Pingree Hill Publishing, Auburn, NH, 2023).

Cosmos Theory: The Sub-Particle Gambol Model (Pingree Hill Publishing, Auburn, NH, 2023).

Cosmos-Universe-Particle-Gambol Theory (Pingree Hill Publishing, Auburn, NH, 2024).

Fractal Cosmos Curve: Tensor-based Cosmos Theory (Pingree Hill Publishing, Auburn, NH, 2024).

The Eternal Form of Cosmos Theory Third Edition (Pingree Hill Publishing, Auburn, NH, 2024).

Fundamental Constants of Cosmos Theory and The Standard Model (Pingree Hill Publishing, Auburn, NH, 2024).

Geometric Cosmos Geometric Universe (Pingree Hill Publishing, Auburn, NH, 2024).

Structure and Dynamics of Cosmos Theory and the Unified SuperStandard Theory (Pingree Hill Publishing, Auburn, NH, 2024).

Available on Amazon.com, bn.com Amazon.co.uk and other international web sites as well as at better bookstores.

CONTENTS

FIGURES and TABLES

Introduction

It is difficult to describe a Cosmos that has no beginning and no end. This book describes such a Cosmos – not as a sequence in time – but rather as an ongoing reality that embodies certain concepts that reveal its features. The author's purpose was to reduce the understanding of Physical reality to the simplest possible terms and yet provide a complete, compact description - backstage simplicity behind the play on stage.

As a preliminary, this author again shows the Electromagnetic Fine Structure Constant 1/137… is exactly determined for each point in the universe, and points out the velocity of light is variable - dependent on gravitation in the vicinity of Black and White Holes. *These results may have a great significance for theories of the universe's structure.*

Reality begins with a mathematical point, and through a series of fractal steps, generates a 2 dimension square that grows to be a 2 dimension space. This space has a simple scaling dynamical equation with a 2 dimension "energy" spectrum. The spectrum can be viewed as a set of four sequences of powers of certain numbers.

One sequence is for the set of spaces of Cosmos Theory. This sequence defines the structure of the Cosmos prior, and separately, from any universes. It exists forever since there is no time before universes are created. Time exists only in universes.

Another sequence is for the 4 known (and any other) forces (groups and interactions). The Standard Model interactions within our universe have this sequence of interactions.

The third and fourth sequences are for the two types of fermion mass sequences (previously described by this author). These sequences approximate the known fermion masses.

There is a fifth sequence that tracks the set of four sequences. This sequence reflects the four above sequences. It suggests fundamental fermions come in fours. Each fermion is one of a quartet. We treat the sets of four fermions as strata. Our universe has 256 fermions in the Unified SuperStandard Theory of the author. They become four strata with a total of 1024 fundamental fermions in Sequence Theory. We see 3 neutrinos in the first stratum. We propose the other known fermions are in the second stratum. The masses of the fermions in the various strata differ by many orders of magnitude. The third and fourth strata are unknown at present.

The existence of strata and the small masses of the first stratum fermions enable us to identify them as confined gambols within the known fermions of the second stratum. The result is the author's Gambol Theory. Gambols are now concrete although they are confined at present energies.

This book ties together Cosmos Theory and Gambol Theory in a compact, consistent fabric for the Cosmos as we know it. A major leap forward starting from a point beyond time! A profound view of the backstage of Reality.

1. QED Eigenvalue Function for the Fine Structure Constant α

The importance of the Fine Structure Constant α is particularly apparent in Astrophysics where it appears in Standard Models of universe evolution. If this constant is fixed at short distances, as we suggest, then Standard Models cannot use a varying value for α.

In 1973 the author developed an approximation to the Eigenvalue Function for the Fine Structure Constant α in the Johnson-Baker-Willey[1,2] (JBW) model of Quantum Electrodynamics that has withstood the passage of time. In 2019 a careful study of the author's Eigenvalue Function calculation revealed a zero at exactly the place of the known value of α.

The author's approximate calculation of the eigenvalue function was to all orders in perturbation theory as opposed to other attempts to determine coupling constants based on extrapolations of fourth order perturbation theory. The author's calculation reproduced the first three known low order perturbation theory calculated values exactly. In 2019 the author found[3]

We have examined the values of the quantities in eq. 1.8 looking for an essential singularity (eq. 1.1) or its approximation. Fig. 1.2 below plots $F_2(\alpha)$ as a function of g.[4] It displays a "flat region." While essential singularities usually are thought to imply a transcendental function such as $\exp(1/\alpha)$, a constant function with value zero fulfills the essential singularity conditions. Therefore we take the "flat region" to indicate an essential singularity.

Fig. 1.2 shows a "close up" of the flat region[5] where F_2 is approximately zero. Upon close numeric analysis[6] we find the results in Fig. 1.1.

g =	-0.0005805369 0000	-0.0005805369 1948	-0.0005805369 5000
α =	0.007297352	*0.0072973525693*	0.007297353
$F_2 \times 10^{10}$ =	3.26316 06817671	3.26316 025452474	3.26316 134861337

Figure 1.1. Values of g, α and $F_2(\alpha) \times 1010$. F_2 is very close to zero for the displayed range of values and throughout the flat region. F_2 has a local minimum at precisely the known value of α = 0.0072973525693(11) = 1/137.0359991(21) (modulo possible experimental uncertainties).

---

[1] K. Johnson, R. Willey and M. Baker, Phys. Rev. **163**, 1699 (1967) and references therein.
[2] Stephen L. Adler, Phys. Rev. D **5**, 3021 (1972) and references therein.
[3] Blaha (2019f).
[4] We describe the F_2 eigenvalue function later.
[5] These figures appeared in Blaha (2019a) and (2019b). We renumbered figures here for clarity.
[6] Since α is experimentally known to 13 places we must use double precision mathematics for these calculations. We used Microsoft Excel.

Thus we have a very good approximation $F_2(\alpha) \cong 0$ at the experimentally known value that is exact to 13 places with a minimum in $F_2(\alpha)$ as anticipated.

F_2 is nearly zero, as are its derivatives, at the physical Fine Structure Constant. It closely approximates a trivial essential singularity of constant value zero in a neighborhood of the singularity.

Note $F_2(\alpha = 0) = 0$ as well. This zero can be viewed as a type of singularity. If QED could transition from positive α to negative α then it would lead to a catastrophe since like charges would then attract.[7,8]

It is extremely important to note the calculation is strictly QED. Thus α is space and time independent, and not Anthropic.

The value of α is constant based on QED. It is locally calculated. General Relativity is based on space-time being locally flat in the small. Therefore α has the same value as calculated in QED at every point in space and time. It is constant throughout the universe for all time.

The negativity of g, the power of p/Λ, at the value of α confirms the finiteness and reasonableness of the calculated Eigenvalue Function.

$g = -0.000580537, \alpha_{calculated}(g) = 0.0072973525693(11) = 1/137.0359991(12)$ (modulo experimental uncertainties)

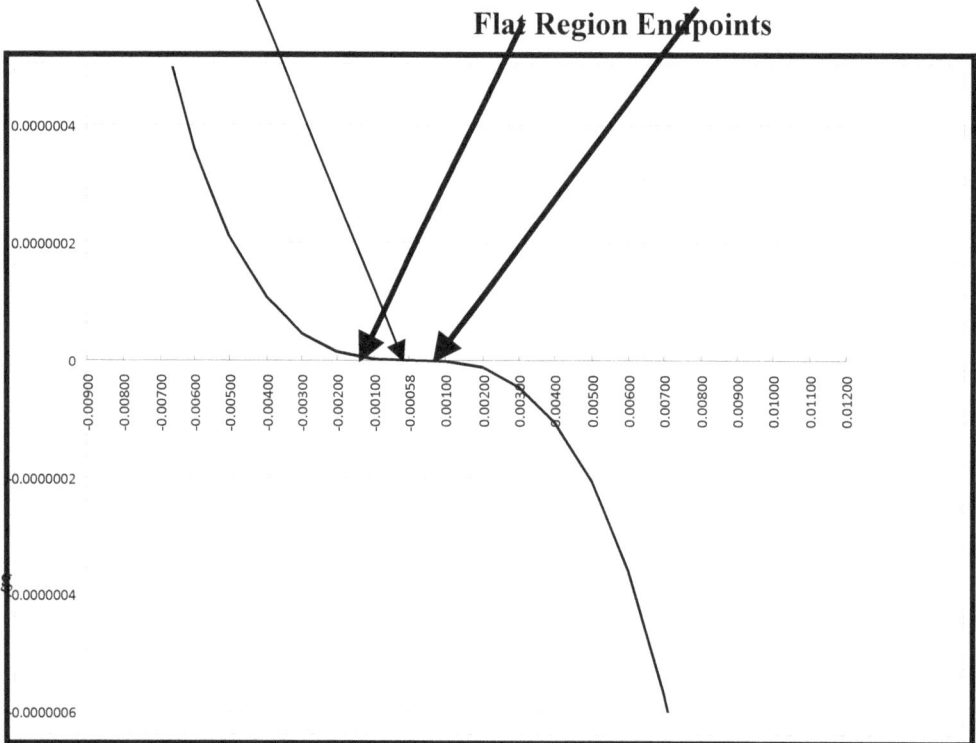

Figure 1.2. Close up plot of our eigenvalue function $F_2(g)$ (vertical axis) vs. g. See section 1.2 for a discussion of $F_2(g)$.

---

[7] Freeman Dyson has speculated on this possibility.
[8] $F_2(\alpha)$ may have more than one zero. One of the zeroes is at the value of the Fine Structure Constant as we show.

1.1 Eigenvalue Function Background

In 1974 this author[9] formulated an approximation to the equations of massless Johnson-Baker-Willey QED and solved them for the vacuum polarization, electron self-energy and the vertex renormalization.

The approximate solution for $F_1(\alpha)$ had the encouraging feature that it *exactly* reproduced the known[10] low order exact calculations of $F_1(\alpha)$:

$$F_{1\text{ low order}}(\alpha) = 2/3 + \alpha/(2\pi) - (1/4)[\ \alpha/(2\pi)]^2 \tag{1.1}$$

Our approximate solution, which summed pieces of the vacuum polarization, yielded the algebraic equations:[11]

$$A_1 = (g + 1)(1 - 2g^2)/[(g + 2)(g - 1)] \tag{1.2}$$

$$A_2 = [8g^2(2g + 1) - (2g^3 + 2g^2 + g - 2)(g^2 + 2g + 2)]/[2(g^2 - 1)(g^2 - 4)]$$

$$A_3 = -2(1 + 3g + 6g^2 + 2g^3)/[g(g + 1)]$$

$$A_4 = -(g + 2)(1 + 5g + 6g^2 + 2g^3)/[g(g^2 - 1)] - 1/(g + 1)$$

$$\psi = [gA_3 - (4 + 2g)A_1]/[(4 + 2g)A_2 - g\,A_4]$$

$$(\alpha/2\pi) = [gA_4 - (4 + 2g)A_2]/(A_4A_1 - A_2A_3)$$

$$F_1(g) = (2/3)(1 - 3g^2/2 - g^3) - (\alpha/4\pi)[(2 + 4g + 4g^2)(g - 2) + \psi g^3]/[(g^2 - 1)(g - 2) + \\ + (\alpha/4\pi)(2 + 4g + 4g^2)(g - 2) + (\alpha/4\pi)\psi g^3]$$

as a function[12] of g with ψ specifying the gauge,[13] and with the definitions

$$\Gamma_\mu(p) = f(\gamma_\mu + 2g\gamma \cdot pp_\mu/p^2)(p/\Lambda)^{2g} \tag{1.3}$$
$$S_F = [f\gamma \cdot p(p/\Lambda)^{2g}]^{-1} \tag{1.4}$$
$$\Gamma_{\mu\alpha}(p) = (f_3/p^2)(\gamma \cdot p\gamma_\mu\gamma_\alpha - \gamma_\alpha\gamma_\mu\gamma \cdot p)(p/\Lambda)^{2g} \tag{1.5}$$

and

$$F_1(g) = (2/3)(1 - 3g^2/2 - g^3) - f_3/f \tag{1.6}$$

[9] Stephen Blaha, Phys. Rev. **D9**, 2246 (1974).
[10] J. Rosner, Phys. Rev. Lett. **17**, 1190 (1966).
[11] Blaha *op. cit.* The solution for the eigenvalue function is clearly best expressed in terms of the g factor in the exponents of the divergent renormalization factors. The form of the equations allows them to be solved as a function of g in the order A_1, A_2, A_3, A_4, ψ, α and F_1.
[12] We use $F_1(g)$ and $F_1(\alpha(g))$ interchangeably.
[13] The JBW massless QED model is gauge invariant. We have opted to choose a gauge using ψ. A choice of gauge is allowed. It makes our approximation fully consistent. We note that some previous renormalization calculations have been based on a choice of gauge.

in the notation of our 1974 paper.[14] (Note g is the power of (p/Λ) where Λ is the cutoff.) The above equations manifestly cannot lead to a form of F_1 with an essential singularity due to their algebraic form.

The plot of F_1 below did not show a zero of F_1 at the physical fine structure constant (See Fig. 1.3 below from Blaha (2019a.) Thus the hopes raised by the JBW QED model seemed dashed—at least in our approximate solution in 1974. In 2019 we revived the hope of a satisfactory eigenvalue function with an eigenvalue at the physical value of the fine structure constant α.

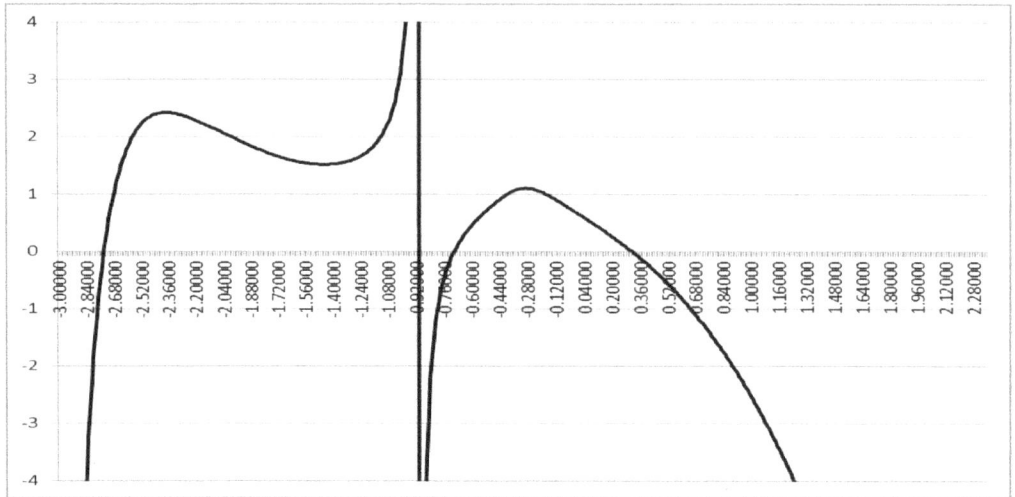

Figure 1.3. A plot of the approximate eigenvalue function $F_1(g)$ (vertical axis) *as a function of g*. Note none of its zeroes correspond to the known physical value of α. It does not have an essential singularity.

1.2 The Quantities g and the New Eigenvalue Function $F_2(g)$

It appears that $F_1(g)$ is too dependent on the first few terms of its perturbation theory expansion. *The essence of the eigenvalue equation, as noted by this author and others,*[15] *is in the asymptotic form of its perturbation theory expansion embodied in eqs. 1.2.*

Consequently we defined a new expression $F_2(\alpha) \equiv F_2(g)$ with the known first three perturbation theory terms removed. The quantities g and $F_2(\alpha)$ are described in Blaha (2019b) and below. We defined $F_2(\alpha)$ with

$$F_2(\alpha) = F_1 - [2/3 + \alpha/(2\pi) - (1/4)[\alpha/(2\pi)]^2] \qquad (1.7)$$

The resulting modified eigenvalue condition led to the success in finding α to its experimentally known values. See Figs. 1.4 and 1.5.

---

[14] Note f_3 is determined by comparing eq. 1.6 and the immediately preceding expression for F_1 in eq. 1.2.
[1515] Stephen L. Adler op. cit. among others.

1.3 Implications

Our calculation of the Fine Structure Constant has several important implications:

1. The value of α has been determined based on the short distance behavior of QED.

2. Since General Relativity is based on the assumption that space-time is locally flat everywhere (except isolated singular points), our QED calculation of α applies throughout the universe.

3. Astrophysical Standard Models should be based on a constant value of α throughout the universe.

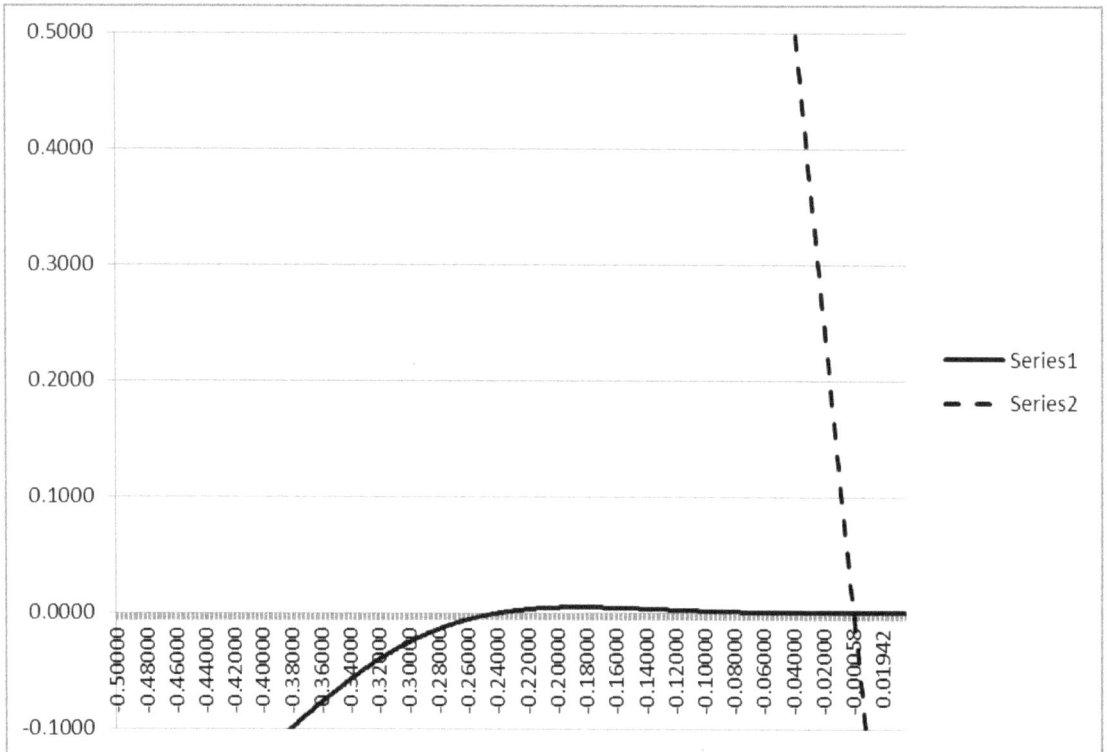

Figure 1.4. Plot of $F_2(g)$ vs. g. The vertical axis displays values of $F_2(g)$. The solid line $F_2(g)$. The dashed line is $\alpha(g)$.

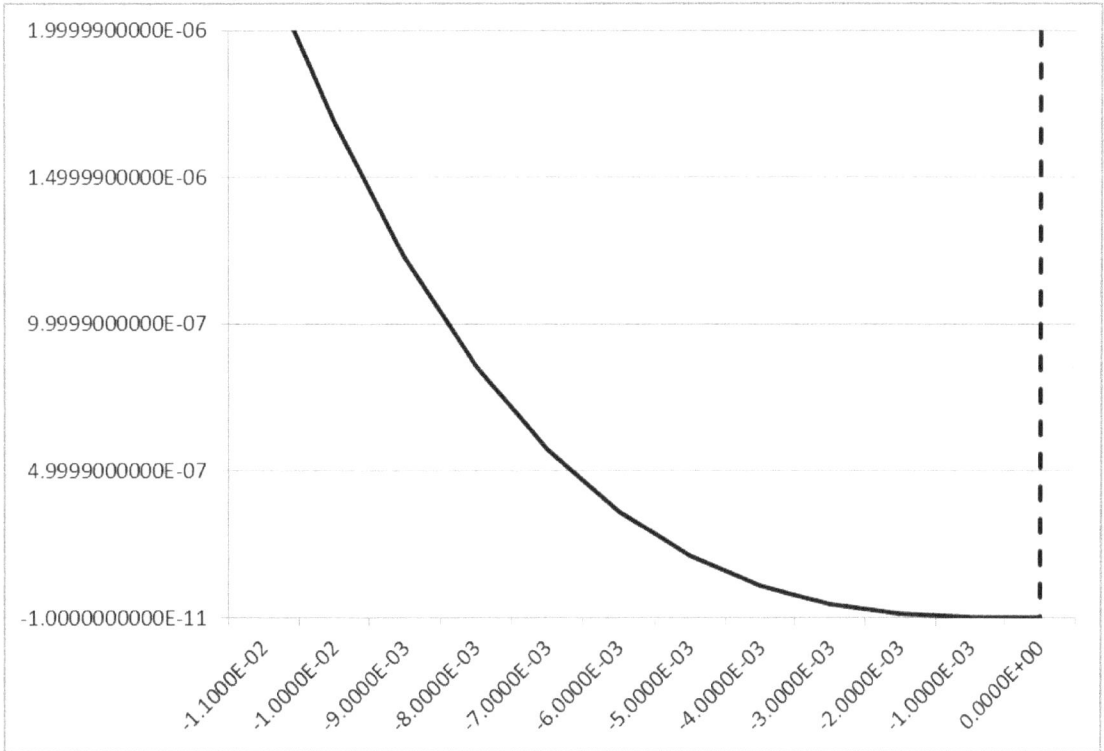

Figure 1.5. Closeup of plot of $F_2(g)$ vs. g. The vertical axis displays values of $F_2(g)$. The solid line $F_2(g)$. The dashed line is $\alpha(g)$. Note the value of $F_2(g)$ rises sharply on this scale as g becomes more negative confirming the zero of $F_2(g)$ at the value of α shown at the beginning of this chapter.

2. The Almost Flat Section in the Fine Structure Constant Eigenvalue Function

The plots of Fig. 1.2, and Fig. 2.1 below, show a "flat" region in the vicinity of the value of g corresponding to the exact value of α. We have previously suggested that eigenvalue functions also exist for the other coupling constants. We believe these other eigenvalue functions have a similar form to the QED eigenvalue function. We will show their values appear in the flat region of the QED eigenvalue function. Common flat regions of all coupling constants reflect the unity of their origin within the framework of Cosmos Theory. This unity is reflected in the coupling constant sequence that we introduced previously and extend later in this book. Fig. 2.1 displays a close-up of the almost flat region in F_2.

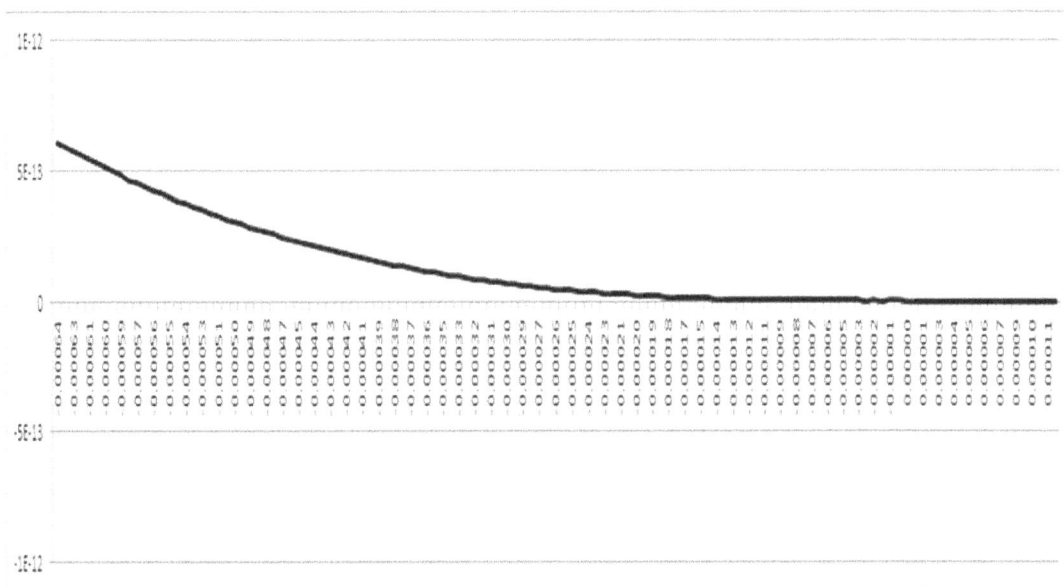

Figure 2.1. Very closeup of plot of $F_2(g)$ vs. g. The vertical axis displays values of $F_2(g)$. Note the value of $F_2(g)$ rises gradually on this scale as g becomes more negative confirming the gentleness of the rise of the almost flat region.

2.1 Extension of Eigenvalue Function Form

The form of the QED eigenvalue function was modified to

$$F_2(\alpha) = F_1 - [2/3 + \alpha/(2\pi) - (1/4)[\alpha/(2\pi)]^2]$$ (1.7)

with the first few perturbation theory terms[16] subtracted based on the reasoning:

It appears that $F_1(g)$ is too dependent on the first few terms of its perturbation theory expansion. *The essence of the eigenvalue equation as noted by this author and others[17] is in the asymptotic form of its perturbation theory expansion embodied in eqs. 1.2.*

After some study of the forms and numeric of $F_2(\alpha)$ we find that a further subtraction appears reasonable. Based on the form of the three known perturbation theory terms we found a modified QED eigenvalue condition $F_\alpha(\alpha)$ makes sense:[18]

$$F_\alpha(\alpha) = F_1 - [2/3 + \alpha/(2\pi) - (1/4)(\alpha/(2\pi))^2 + \mathbf{1/5}\,(\alpha/(2\pi))^3] \qquad (2.1)$$

where we choose the coefficient 1/5 due to its similarity to the previous three known coefficients. The corresponding coefficient vacuum polarization term is not known in QED perturbation theory and apparently intractable.

   The resulting F_α has a slight modification of the flat region compared to F_2 – making it flatter. Fig. 2.2 plots F_α vs. g in the flat region for the four known coupling constants listed in Fig. 3.2 below. The values of F_α range from 10^{-11} to $5 \cdot 10^{-8}$.

   Fig. 2.3 plots F_α vs. g in the flat region for the SU(4) α_4 coupling constant listed in Fig. 3.2 below. The value of F_α is 10^{-6}.

   The close approximation of the F_2 results and the F_α results shows the resilience of our calculation of the Fine Structure Constant. The coefficient of the $(\alpha/(2\pi))^3$ can be expected to be less than the preceding terms' coefficients for a convergent Eigenvalue Function unless the function has an essential singularity. Thus the choice of 1/5. (Our F_1, F_2, and F_α functions do not have essential singularities.)

   The close approximation of F_α coupling constant values to known coupling constants may well reflect the interrelations of the coupling constants that we discuss later.

2.2. Vacuum Polarization Structure and the g Parameter
   The g parameter appears in the author's calculation of the Eigenvalue Function vacuum polarization in momentum space:

$$(p/\Lambda)^{2g} \qquad (2.2)$$

where p is the momentum and Λ is the cutoff. Note that the values of g for the Fine Structure Constant and other coupling constants in Fig. 2.2 are negative as they should be since the calculation is not divergent.

   In chapter 3 we show the coupling constants form a sequence governed by a master equation for Cosmos sequences eq. 3.1. Thus there is some reason to believe that

---

[16] These terms appear exactly in Blaha's calculation of the QED eigenvalue equation: Stephen Blaha, Phys. Rev. **9**, 2246 (1974).

[17] Stephen L. Adler op. cit. among others.

[18] We use the subscript α to avoid confusion with Blaha's f_3 function in his paper's calculation.

the vacuum polarizations of the vector bosons embodying interactions are related. The trend of the g values in Fig. 2.2 parallels the trend of interaction strengths in Fig. 3.2.

The powers of momenta in eq. 2.2 show a sharper (steeper) cutoff in momentum as g becomes increasingly negative. The coordinate space shape of the vacuum polarization therefore becomes broader with increasingly negative g. As a result we see the vacuum polarizations of interactions become broader in coordinate space with the growth of interaction strength. These accords with physical expectations.

2.3. Universe Standard Models in view of Constant α and Variable Speed of Light

There are a number of attempts to create a Standard Model for Universes. These models necessarily depend on α and the speed of light. We have found that α, and other coupling constants, are fixed and not variable. *Since α depends on local physics in approximately flat patches and since all regions of the universe must support this locality for all times, we find α to be constant throughout the universe for all times. It cannot be variable!*

In Blaha (2025a) we have found that the speed of light depends on the gravitation of bodies.

Together these properties form a significant constraint on Standard Models of the universe.

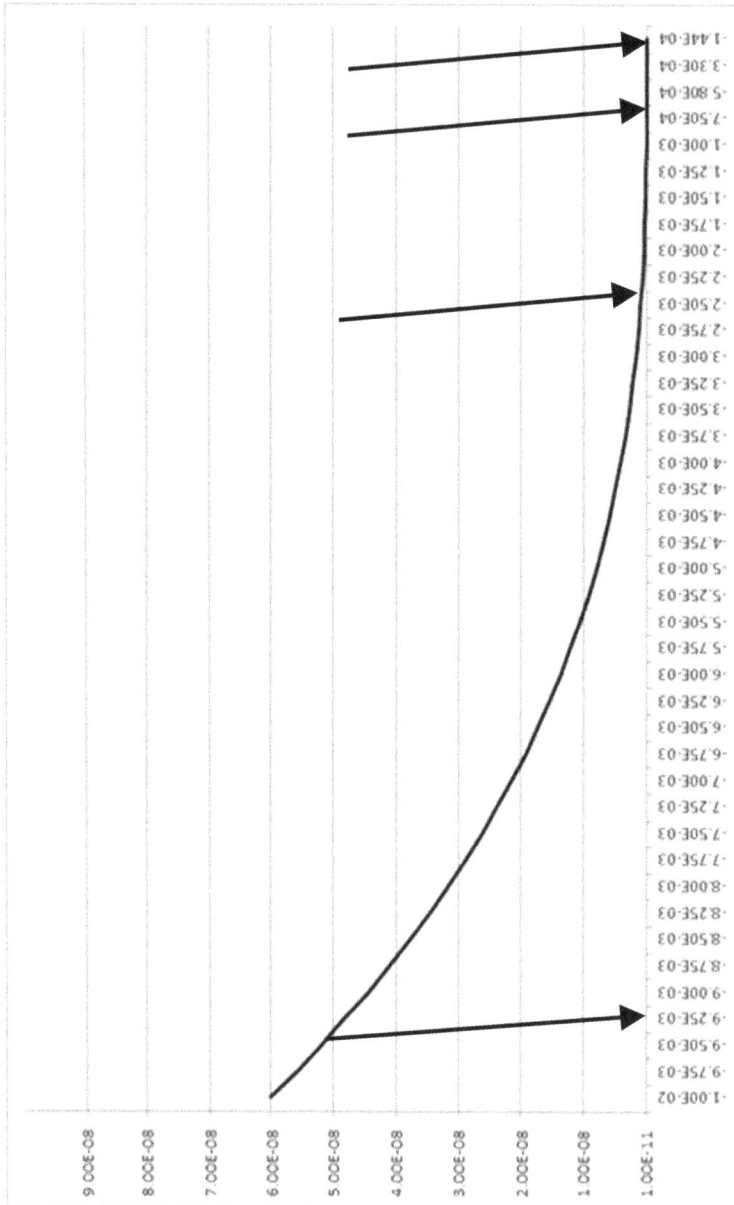

Figure 2.2. Very closeup of plot of $F_\alpha(g)$ vs. g. The vertical axis displays values of $F_\alpha(g)$. Note the value of $F_\alpha(g)$ rises gradually on this scale as g becomes more negative confirming the gentleness of the rise of the almost flat region. The g values are: -000144, -0.00058, -.0023, -0.00925 for the α_0, α_1, α_2, α_3 coupling constants respectively from Fig. 3.2. The arrows indicate the coupling constants on the plot.

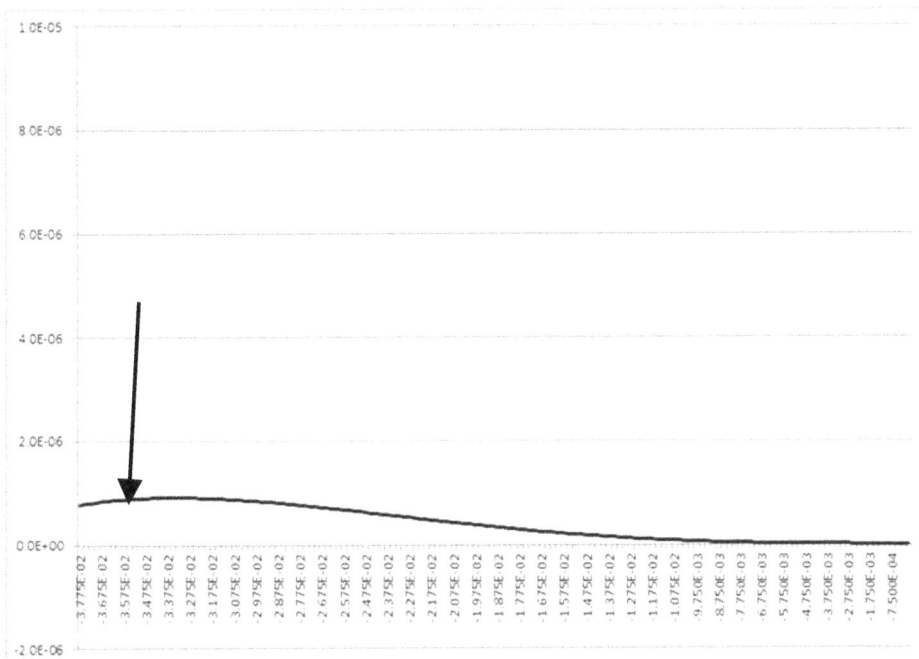

Figure 2.3. Closeup of plot of $F_\alpha(g)$ vs. g. The g value for the α_4 coupling constant is g = -0.03575. See Fig. 3.2. The arrow indicates the coupling constant on the plot.

3. Stability of Cosmos Structure

3.1 The Timeless Cosmos Structure

The structure of the Cosmos is specified by a set Cosmos spaces. See Figs. 3.1a and 3.1b for the set of spaces (not including HyperCosmos spaces of the Second Kind or Limos spaces that we describe in previous books).

Since the structure of the Cosmos spectrum of spaces does not have substance, its set of spaces may be viewed as a set of dimension arrays. The spectrum of spaces is a conceptual definition without a need for the specification of an origin. In that respect it is analogous to Plato's Theory of Ideals.

The set of Cosmos spaces is timeless and unchangeable. It exists prior to any definition of time. It is beyond any definition of time. The definition of a time is a feature of a universe. Cosmos spaces are independent of universes. They are essentially a blueprint of the Cosmos. The origin of Cosmos structure is beyond Physics in the author's view. The form of the structure is sequential – based on powers of 2. We attribute the sequential structure to a new theory – presented here for the first time – that develops the form of spaces and universes through a set of four types of sequences.

The universes that implement a space definition are created entities. We have shown in previous books that an initial parent universe(s) must exist that contains (possibly chains of) sub-universes. The origin of a parent universe is problematic. We can say, as we have previously, that the parent universe(s) must satisfy a certain energy-like consistency condition for it to exist. Beyond that, we can only view its origin as beyond our current endeavors. The dynamics that generates/creates sub-universes also remains to be thoroughly investigated.

We view the "creation" of the Parent universe as the beginning of a *time*. (There can be no time (or dynamics) prior to the existence of a space-time. A space-time only exists within a universe. Thus time begins within a parent universe at its creation point.) The times of the sub-universes, including their creation times, are relative to the parent universe time. "No time before time."

3.2 Primal Cosmos Universe

We have shown in previous books that sub-universes may be created within a parent universe based on consistency condition or based on a Quantum Field Theory of universes. Sub-universes have substance – energy, mass and interactions. Sub-universes would appear to be created as a point that subsequently expands (Big Bang) to a possibly growing space-time.

3.3 The Sequence Theory Equation and Cosmos Theory

The set of spaces of Cosmos Theory has a sequential form based on powers of 2. We will see it is one of a set of four types of sequences of our new theory, *Sequence Theory*. We base Sequence Theory on a fundamental equation, somewhat like the Schrödinger equation but with a scaling form – independent of dimension.

The sequence equation base of Cosmos Theory must exist timelessly. It specifies the structure of the Cosmos – its spectrum of spaces and their contents, which are sets of dimensions.

The role of the fundamental sequence equation is as a description of the Cosmos. It is not Reality. It is a mathematical reflection of Reality that identifies the steps connecting the parts of Reality.

We have found what may be the core equation of Cosmos Theory that connects the spectrum of Cosmos Spaces, and the features of fermions and interactions in our universe – the only universe with which we are familiar. The core equation is purposefully simple.

It has a basis in a two dimension space with no time variable. We view this space as originating in a mathematical point. Through a sequence of fractal-like line segment increments it becomes a one dimension line. Then through a further sequence of increments it undergoes fractal "transformation" to a unit square of two dimensions. Under scaling, the unit square becomes an infinite two dimension space. We call the resulting space the Fractal Cosmic Curve space.[19] The Cosmic Curve is similar in construction to the Hilbert fractal curve.

We specify it by a simple Schrödinger-like equation in coordinates x and y with a λ/r^2 scaling potential. The scaling potential provides the simplest guarantee of an energy spectrum that is a sequence of powers of a constant.

$$(\partial^2/\partial x^2 + \partial^2/\partial y^2 + \lambda/r^2 + C)\psi = 0 \tag{3.1}$$

where λ is a coupling constant and $r = (x^2 + y^2)^{\frac{1}{2}}$. In polar coordinates

$$x = r\cos(\varphi)$$
$$y = r\sin(\varphi)$$

we find

$$(\partial^2/\partial r^2 + 1/r\, \partial/\partial r + \partial^2/\partial\varphi^2 + \lambda/r^2 + C)\psi = 0 \tag{3.1a}$$
$$(d^2/d\varphi^2)\psi = 0$$

and the solution

$$\psi = \psi(r)\Phi(\varphi) \tag{3.1b}$$

where

$$\Phi(\varphi) = \exp(ik\varphi) \tag{3.1c}$$

---

[19] This is detailed in Blaha (2024c), *Fractal Cosmic Curve: Tensor-Based CosmosTheory.*

The quantum number k must be a whole integer to avoid irregularities in $\psi(r)$. We consider the case of $k = 0$. Substituting and separating in eq. 3.1a we find

$$(\partial^2/\partial r^2 + 1/r \ \partial/\partial r + \lambda/r^2 + C) \ \psi(r) = 0 \tag{3.1d}$$

The eigenvalue spectrum due to eq. 3.1d is

$$C_n = \text{const} \ \exp(-n\pi/(\lambda - \tfrac{1}{4})^{\frac{1}{2}}) \tag{3.2}$$

based on Case's paper.[20] The quantization of the C_n eigenvalue spectrum is the requirement of orthogonality of solutions based on an observation of Von Neumann.

Thus we initially have a radial energy series in eq. 3.2. We will generalize this series to a double series in n and an angular momentum quantum number n_φ described in section 3.4 based on a Bohr-like semiclassical atomic model:

$$C_{n_\varphi n} = c_{n_\varphi} \ \exp(n\pi/(\lambda_{n_\varphi} - \tfrac{1}{4})^{\frac{1}{2}}) \tag{3.2a}$$

where n is the radial quantum number and n_φ is the angular momentum quantum number. We take the Cosmos radial sequences parameterized by n_φ to have the form:

$$C_{n_\varphi n} = c_{n_\varphi} \ \exp(n\pi/(\lambda_{n_\varphi} - \tfrac{1}{4})^{\frac{1}{2}}) \tag{3.2b}$$

giving rising values with n. The growth factor for each power series sequence is

$$g_{n_\varphi} = \exp(\pi/(\lambda_{n_\varphi} - \tfrac{1}{4})^{\frac{1}{2}}) \tag{3.2c}$$

The radial quantum number n parameterizes each sequence of "energy" values. The angular momentum quantum number n_φ, which we define in the next section, specifies the sequence and its constant factor c_{n_φ}. The index n_φ determines the coupling constant $\lambda = \lambda_{Bohr \ n_\varphi}$ in eq. 3.1, for each specific sequence.

3.4 Bohr View of Sequence Equation

Bohr[21] developed an atomic model which successfully initiated quantum theory. We use it to illustrate the physical origin of the angular momentum quantization in n_φ. We now adapt that model to the case of eq. 3.1 transformed to a classical two dimension equivalent with attractive potential λ/r^2. Assuming the existence of stationary, discrete states in the 2-dimension model "atom" of eq. 3.1a with an atomic like nucleus we assume an "electron" mass m, velocity v and radius r. The centripetal force equals the force associated with the potential:

$$mv^2/r = 2\lambda/r^3 \tag{3.3}$$

---

[20] K. M. Case, Phys. Rev. **80**, 797 (1950).
[21] N. Bohr, Phil. Mag. **26**, 1 (1913).

The kinetic energy is

$$T = \tfrac{1}{2}\,mv^2 = \tfrac{1}{2}\,\lambda/r^2 \tag{3.4}$$

We assume the stationary states have a quantized angular momentum

$$p_\varphi = mvr = n_\varphi \hbar \tag{3.5}$$

implying

$$v = n_\varphi \hbar/mr \tag{3.6}$$

Eq. 3.4 then implies the quantization of λ with the quantum number n_φ

$$\lambda_{\text{Bohr } n_\varphi} = n_\varphi^2 \hbar^2/m = n_\varphi^2/m \tag{3.7}$$

after setting $\hbar = 1$.

This semiclassical result allows the Bohr model radius *not* to be quantized but gives a quantization consistency condition on the coupling constant. This Bohr-like model suggests a quantization of the coupling constant λ that goes beyond the radial quantization of the energy C_n in eq. 3.2 based on the requirement of orthogonality in Case's paper.

While m appears as a mass factor we may view it as a dimensionless constant since λ is a dimensionless constant.

The quantization of the coupling constant λ reflects the $1/r^2$ potential. This feature is not present in the hydrogen Schrödinger and Dirac equations due to their $1/r$ potential.

3.5 First Look at Sequence Theory

Eq. 3.1 defines the basic equation of Sequence Theory. It applies to all sequences. The sequence relations that follow are:

$$\lambda_{\text{Bohr } n_\varphi} = n_\varphi^2/m \tag{3.8a}$$
$$g_{n_\varphi} = \exp(\pi/(\lambda_{\text{Bohr } n_\varphi} - \tfrac{1}{4})^{\frac{1}{2}}) \tag{3.8b}$$
$$\lambda_{\text{Bohr } n_\varphi} = [(\ln(g_{n_\varphi})/\pi]^{-2} + \tfrac{1}{4} \tag{3.8c}$$
$$C_{n_\varphi n} = c_{n_\varphi} \exp(n\pi/(\lambda_{\text{Bohr } n_\varphi} - \tfrac{1}{4})^{\frac{1}{2}}) \tag{3.8d}$$

The sequences are characterized by eqs. 3.2 witch has the initial values of the sequences: c_{n_φ}. This constant c_{n_φ} is dimensionless for sequences of dimensions. It has the dimension of [mass] for sequences of masses. For Cosmos dimension arrays c_{n_φ} is dimensionless and has its value set by the number of creation/annihilation operators for a PseudoQuantum fermion wave function.

This section, and the following sections, shows that the values of the $\lambda_{\text{Bohr } n_\varphi}$ for the power series sequences of the Cosmos Theory spaces spectrum, the coupling constants, and the fundamental fermion mass sequences have a simple relationship based on their order: $n_\varphi = 4, 2, 1, \tfrac{1}{2}^{\frac{1}{2}}$ and $\tfrac{1}{2}$.

The procedure to determine λ_{n_φ} and g_{n_φ} that we follow is:

1. Calculate $\lambda_{Bohr\ n_\varphi}$ from eq. 3.8a.
2. Calculate g_{n_φ} from eq. 3.8b.

THE HYPERCOSMOS SPACES SPECTRUM

Blaha Space Number $N = o_s$	Cayley-Dickson Number n	Cayley Number C_n d_c	Dimension Array column length $d_{cd}=d_{cr}$	Dimension Array Size $d_{dN}=d_{dr}$	Space-time-Dimension r	CASe Group $su(2^{r/2},2^{r/2})$ CASe
0	10	1024	2048	2048^2	18	su(512,512)
1	9	512	1024	1024^2	16	su(256,256)
2	8	256	512	512^2	14	su(128,128)
3	7	128	256	256^2	12	su(64,64)
4	6	64	128	128^2	10	su(32,32)
5	5	32	64	64^2	8	su(16,16)
6	4	16	32	32^2	6	su(8,8)
7	**3**	**8**	**16**	$\mathbf{16^2}$	**4**	**su(4,4)**
8	2	4	8	8^2	2	su(2,2)
9	1	2	4	4^2	0	su(1,1)
10	0	1	2	2^2	-2	

Figure 3.1a. The Cosmos Theory HyperCosmos space spectrum. Note: $n_\varphi = 4$. The parameter n is the Cayley-Dickson number. See Blaha (2022c).

Blaha-22   The Sequence Theory Foundation of Cosmos Theory and Gambol Theory

Level

1   88 Dimension
     UltraUnification
          Space

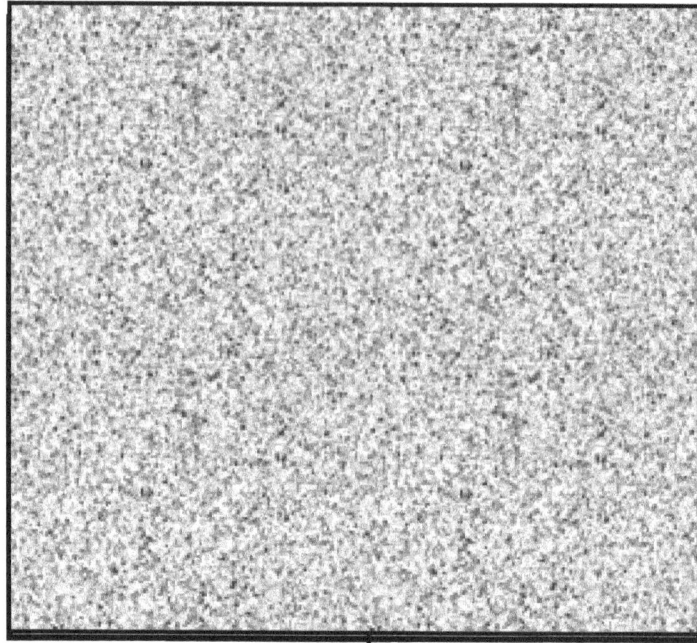

     42 Dimension
Full HyperUnification
          Space

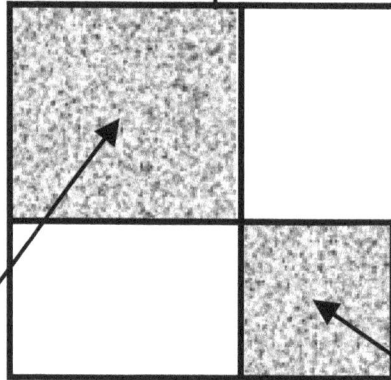

2

10 HyperCosmos HyperUnification Spaces

10 Second Kind
HyperCosmos HyperUnification Spaces

3

10 HyperCosmos Spaces

10 Second Kind HyperCosmos Spaces

4

Figure 3.1b. Diagram of the four levels of spaces of Cosmos Theory. It contains 42 spaces. From Blaha (2023a).

3.5.1 λ Value for Cosmos Theory Power Sequence of Spaces

The Cosmos Theory set of spaces has a λ value for the doubling of the length of a dimension array column.[22] (See Fig. 3.1)

$$d_{cd} = 2^{r/2 + 2} \tag{3.9}$$

For $n_\varphi = 4$ we set

$$2 = \exp(\pi/(\lambda_4 - \tfrac{1}{4})^{\frac{1}{2}})$$

Solving for λ_4 we obtain

$$\lambda_{\text{Bohr } n_\varphi=4} = 16/m = \lambda_4 = 20.79 \tag{3.10}$$

for $n_\varphi = 4$. As a result, eq. 3.8a implies

$$m = 16/\lambda_{\text{Bohr } n_\varphi=4} = 0.7696 \tag{3.11}$$

and

$$m \approx \pi/4 = 0.785 \tag{3.12}$$

We will use the value of $m = 0.7696$ in the cases below. Note m is dimensionless by eq. 3.7. In general eq. 3.7 specifies λ_{n_φ}.

3.5.2 λ Value for the Coupling Constant Power Sequence

The Coupling Constant series has a multiplicative power of four embedded in the known coupling constants. (See Fig. 3.2, which we presented in previous books.)

We can map the number of degrees of freedom to the coupling constant spectrum for the couplings α_i by setting

$$4 = \exp(\pi/(\lambda_{\text{Bohr } n_\varphi=2} - \tfrac{1}{4})^{\frac{1}{2}}) \tag{3.13}$$

to generate the Cosmos coupling constant spectrum of Fig. 3.2. As a result eq. 3.13 gives

$$\lambda_{\text{Bohr } n_\varphi=2} = [(\ln(4)/\pi]^{-2} + \tfrac{1}{4} = 5.386 \tag{3.14}$$

However, eq. 3.8a gives

$$\lambda_{\text{Bohr } n_\varphi=2} = 4/m = 5.198 \tag{3.15}$$

Comparing eqs. 3.14 and 3.15 we see the value of m is a good approximation for $\lambda_{\text{Bohr} n_\varphi=2}$ values to 3%. Thus we account for the factor of four between adjoining coupling constant values using the $1/r^2$ potential seen above. See Fig. 3.2.

---

[22] The fractal Cosmos Curve construction is based on column lengths. See Blaha (2024b) and (2024c).

	E S T I M A T E			**E X P E R I M E N T**		
Interaction	**Expression**	$\alpha_i = g^2/4\pi$ **Value**	**g**	**Known**[23] **Value** $g^2/4\pi$	**Deviation**	
$U(0)$ α_0	$e^2/4096$	0.0018036	-	-	-	
$U(1)$ $\alpha_1 = \alpha$	$e^2/1024$	0.00721438	0.303	0.0072973525643	1.15%	
$SU(2)$ $\alpha_2 = g^2/4\pi$	$e^2/256$	0.0289	0.63	0.0316	9.3%	
$SU(3)$ $\alpha_3 = \alpha_S$	$e^2/64$	0.115	1.21	0.117	1.7%	
$SU(4)$[24] α_4	$e^2/16$	0.462	2.4?	0.458	0.087%	

Figure 3.2. Coupling constants for the UST (and the Standard Model). Note factor of four relatio between adjoining α values.

3.5.3 λ Value for Fermion Mass Power Sequences

The approximate masses for the two fermion sequences appear in Fig. 3.3.

	SEQUENCE 1				**SEQUENCE 2**			
	e	u	c	t	v'	d	s	b
Mass:	0.511×10^{-3}	1.80×10^{-3}	1.28	171	8.4×10^{-5}	4.24×10^{-3}	102×10^{-3}	4.34
Multiplier:		$2^5\pi$	$2^5\pi$	$2^5\pi$		32	32	32

Figure 3.3. The fundamental fermion mass sequences with approximate multiplier factors between the masses of each sequence. The sequence 1 masses show a significant variation in their $2^5\pi$ factor but their overall trend is approximately satisfactory. Sequence 2 masses approximate the factor of 32 progression.

Taking the fermion progression factors seriously we can calculate the λ values, which relate to n_φ, for each series:

$$\lambda_{\text{Bohr } n\varphi\text{seq1}} = [(\ln(32\pi)/\pi]^{-2} + \tfrac{1}{4} = 0.714 \qquad \approx g_2 = 0.63 \qquad (3.16)$$
$$\lambda_{\text{Bohr } n\varphi\text{seq2}} = [(\ln(32)/\pi]^{-2} + \tfrac{1}{4} = 1.07 \qquad \approx g_3 = 1.21 \qquad (3.17)$$

where the coupling constants, g_2 and g_3, shown for comparison are taken from Fig. 3.2. Note the similarity of coupling constant values to the λ values. We describe the basis of this similarity in chapter 6.

Eq. 3.8a gives the values

$$\lambda_{\text{Bohr } n\varphi=\text{seq1}} = n_{\varphi1}^2/m \qquad\qquad (3.18)$$
$$\lambda_{\text{Bohr } n\varphi=\text{seq2}} = n_{\varphi2}^2/m \qquad\qquad (3.19)$$

If we set $n_{\varphi2} = 1$ and $n_{\varphi1} = \tfrac{1}{2}^{\frac{1}{2}}$ we find

$$\lambda_{\text{Bohr } n\varphi=\text{seq1}} = n_{\varphi1}^2/m = 0.65 \qquad\qquad (3.20)$$
$$\lambda_{\text{Bohr } n\varphi=\text{seq2}} = n_{\varphi2}^2/m = 1.3 \qquad\qquad (3.21)$$

---

[23] All coupling constant values are based on data from Particle Data Group Tables of 2024.
[24] This value is based on the "doubling trend" seen in the three known coupling constants g_i above in Fig. 3.2.

Thus there is an approximate match between fermion mass sequence factors and coupling constants:

$$\lambda_{\text{Bohr } n_\varphi=\text{seq}1} = 0.65 \approx g_2 = 0.63 \tag{3.22}$$
$$\lambda_{\text{Bohr } n_\varphi=\text{seq}2} = 1.3 \ \approx g_3 = 1.21 \tag{3.23}$$

These values are approximately correct to 3% for sequence 1 and 7.4% for sequence 2. See chapter 6 for a detailed description.

The multiplicative factor for sequence 1 is

$$g_{\text{seq}1} = \exp(\pi/(\lambda_{\text{Bohr } n_\varphi=\text{seq}1} - \tfrac{1}{4})^{\frac{1}{2}}) = 143.64 \tag{3.24}$$
$$= 1.43 \times 32\pi \approx 3/2 \times 32\pi$$

and for sequence 2 is

$$g_{\text{seq}2} = \exp(\pi/(\lambda_{\text{Bohr } n_\varphi=\text{seq}2} - \tfrac{1}{4})^{\frac{1}{2}}) \ = 21.45 \tag{3.25}$$
$$= 2/3 \times 32$$

thus matching the fermion mass growth factors fairly well in Fig. 3.3. The factors of 3/2 and 2/3 are suggestive but their understanding must await a deeper analysis.

3.6 A Clear Progression of n_φ Values

The preceding discussions show an interrelation of the sequences of Cosmos spaces, of coupling constants, and of fermion masses. Thus the model in eq. 3.1 that we constructed for the structure of Cosmos Theory successfully connects these disparate parts together in an overall framework. Fig. 3.4 strongly supports the consistency of the various sequences that we have analyzed. *Their characterization using the n_φ index values shows the underlying connection of the sequences within the framework of eq. 3.1 and the fractal space of its equation. The relation of the n_φ index and $\lambda_{g_{n\varphi}n_\varphi}$ to the multiplier factors of sequences is an important support for Sequence Theory.*

A secondary sequence appears in line E in Fig. 3.4. This sequence is the sequence of n_φ values. *It is nicely sequential.* We discuss this sequence in section 3.7.

Sequence	n_φ	g_g multiplier factor	$\lambda_{\text{Bohr } n_\varphi}$
A Cosmos dimension array column lengths $2^{r/2+2}$	4	2	20.79
B Coupling Constants Multiplier	2	4	5.39
C Fermion Sequence 2 Masses	1	$2/3 \times 2^5$	1.3
D Fermion Sequence 1 Masses	$\tfrac{1}{2}^{\frac{1}{2}}$	$3/2 \times \pi 2^5$	0.65
E n_φ Values Sequence; A Sequence of Sequences	$\tfrac{1}{2}$	95939	0.325

Figure 3.4. Interelated Cosmos Theory Sequences. Note: The fermion Mass sequences must apply to mass sequences in other generations and other layers up to overall factors. The coupling constant sequence must apply to the coupling constant sequences in other layers up to overall factors. The values of n_φ, g_g, and $\lambda_{\text{Bohr } n_\varphi}$ should apply for all generations and layers. The sole difference is the values of the starting sequence values $c_{n_\varphi n}$ from eq. 3.12. Note $\lambda_{\text{Bohr } n_\varphi} > 0.25$ is required by eq. 3.1 in order to have the bound state energy spectrum. Thus the possible additional line with $n_\varphi = \tfrac{1}{4}$ is ruled out.

3.7 Explanation of Sequences

We now consider the meaning of the sequences in Fig. 3.4. First note the first line A applies to the structure of Cosmos Theory in general. The remaining lines are for individual universes. We view these lines as applicable in all universes of all ten Cosmos spaces. The features of the individual sequences are:

A Cosmos dimension array column sizes $(2^{r/2+2})$ Multiplier $n_\varphi = 4$
Cosmos spaces' dimension array columns grow by factors of 2 as the dimension increases by 2. This sequence directly matches the sequence of line segments in the Hilbert fractal curve construction as well as the sequence of line segments in the fractal Cosmos Curve construction.

B   Coupling Constants Multiplier $n_\varphi = 2$
The coupling constants sequence grows by factors of 4. See Fig. 3.2.

C   Fermion Sequence 2 Masses $n_\varphi = 1$
The sequence 2 fermions grow by factors of 32. See Fig. 3.3.

D   Fermion Sequence 1   Masses $n_\varphi = \tfrac{1}{2}^{\frac{1}{2}} = 2^{\frac{1}{2}}/2 = 0.707107$
The sequence 1 fermions grow by factors of 32π. See Fig. 3.3. This n_φ value has a multiplier of $2^{\frac{1}{2}}$ from the $n_\varphi = 1$ value and also to the $n_\varphi = \tfrac{1}{2}$ value. Thus it is mid-sequence. It has the value $\tfrac{1}{2}$ times $2^{\frac{1}{2}}$ - multiplicatively greater than the sequence value of the next $n_\varphi = \tfrac{1}{2}$ value.

E Sequence of n_φ Values $n_\varphi = 1/2$
This is the sequence of n_φ values in Fig. 3.4. The items in line E are:

$$\lambda_{\text{Bohr } n_\varphi=\frac{1}{2}} = n_\varphi^2/m = \tfrac{1}{4}/m = 0.325$$
$$g_{g\frac{1}{2}} = \exp(\pi/(\lambda_{\text{Bohr } n_\varphi=\frac{1}{2}} - \tfrac{1}{4})^{\frac{1}{2}}) \tag{3.26}$$
$$= 95939$$

The sequence of $n_\varphi = \tfrac{1}{2}$ values has a growth multiplier $g_{g\frac{1}{2}} = 95939$. This extremely large value is, at first glance, somewhat strange. It is much larger than the analogous multipliers for fermions in items C and D. However, the relation of the v' and v_e Neutrino masses in Fig. 3.5 suggests that there may be some new parts to the complete spectrum of fundamental fermions. We see[25]

$$v' = v_e \, e^2(\alpha_0^2)^{-1} \approx \pi e^2 g_{g\frac{1}{2}} v_e \tag{3.27}$$

where

$$g_{g\frac{1}{2}} = 95939 \approx 1/(\pi\alpha_0^2) = 97852 \tag{3.28}$$

---

[25] The choice of the v' mass was made in our earlier book. It was based on making sequence 2 masses approximately follow a power sequence form, which was not possible if we used the v_e mass.

as in row E above using (from Fig. 3.2)

$$\alpha_0 = e^2/4096 = 0.0018036$$

Thus we may extrapolate and find a sequence of neutrino masses based on the mass multiplier of 95939. (The C and D items are also for sequences of fundamental fermions.) It is then reasonable to assume that all fundamental fermions in our universe are "multi-plicated."

We define a number n_s labeling a proposed set of replicates of the 256 fermions of the r = 4 Cosmos universe of UST. One replicate of 256 fundamental fermions that includes v_e can be viewed as the $n_s = 1$ set.[26] The next $n_s = 2$ set has the v' neutrino and 255 additional fermions with each of their masses scaled up from their $n_s = 1$ counterpart by a factor of $\pi e^2 g_{g\frac{1}{2}}$.

Seeing that there are four sequences above in Fig. 3.4 and that they have a lower limit of the $n_\varphi = \frac{1}{2}$ sequence and an upper limit conceptually[27] of the $n_\varphi = 4$ sequence we suggest that there are 4 counterparts of the 256 fermion spectrum in our universe with corresponding results in other universes of other spaces. We call each counterpart a *stratum* and label them as $n_s = 1, 2, 3, 4$. See Fig. 3.6.

The set of strata supports an extension of the fermion Cosmos Theory dimension array. We use them to define a Sequence Theory fermion dimension array of 2^{r+6} dimensions due to a factor of 2^2 from the four stratums. We do not introduce new vector or scalar boson strata for reasons described in chapter 4. This section began the exploration of Sequence Theory.

3.8 A Foundation for Sequence Theory

The Sequence chart in Fig. 3.4 supports the concept of Sequence Theory by displaying the close match of the calculated multiplier g_g values, the sequence of n_s values and the known multiplier values of the sequences.

The sequences found in Cosmos Theory are understandable within the framework of Sequence Theory.

---

[26] We use the index n_s for reasons described in chapter 4. The subscript will be seen to index a "stratum."

[27] Partitioning dimension arrays further is not significant. Also the g_g is close to 1 making it of less interest.

$\nu_e = 3.878 \times 10^{-11}$ GeV/c^2

   Our ν_e mass estimate = 0.03878 ev/c^2 is consistent with known data estimates.

$\nu' = \nu_e\, e^2(\alpha_0^2)^{-1} = 0.03878$ ev/c$^2 \times 4096^2/e^2 = 8.8 \times 10^{-5}$ GeV/c^2 where e = 2.718.

   We scale ν_e by $(\alpha_0^2)^{-1}$. This scaling puts the neutrino mass on the same footing as the other masses.
   It may reflect the need to scale the neutrino mass by the U(0) "symmetry." Later we suggest a
   further need to scale mass values based on the ratio of SU(2) mass effects to SU(3) mass effects, which
   appear to vary as a function of fermion mass.

e = 0.511×10^{-3} GeV/c^2 = electron mass

u 1.76 mev/c2 = 1.76×10^{-3} GeV/c^2

   We scale the u mass by a factor of 4/5 – again to put it on a similar footing as the other masses.
   Note: the u mass is 2.2 +0.5 -0.4 $\times 10^{-3}$ GeV/c^2 The 1.76 value is consistent with the lower bound 1.8.

c 1.27 GeV/c^2

t 172.76 GeV/c^2

d 3.76×10^{-3} GeV/c^2

   We scale the d mass by a factor of 4/5 – again to put it on a similar footing as the other masses.
   The 3.76 value is consistent with the tentative lower bound 4.4.

s 95×10^{-3} GeV/c^2

b 4.18 GeV/c^2

Figure 3.5. Fundamental first generation current fermion masses. Each fundamental fermion symbol represents its mass. This is Fig. 1.4 in *Quark, Lepton, W and Z Masses of Cosmos Theory and The Standard Model* Blaha (2024g).

4. Sequence Theory and Gambol Theory

We view Sequence Theory as an extension of Cosmos Theory. It unites the sequences found in Cosmos Theory in a more fundamental manner through the scaling equation eq. 3.1, which originally appeared as a ProtoCosmos Model possibility a few years ago.

It introduces an extension of the sets of fundamental fermions by a factor of four. We see the known neutrino ν_e is in the first of the four strata and the heavier neutrino ν' that we proposed in previous books appears in the second stratum *together with the other known fundamental fermions*. The ν_e, ν_μ, and ν_τ neutrinos are the only known fermions in the first stratum. In our universe (using UST) each of the 256 fundamental fermions, with which we are familiar, has a counterpart in each of the other threes strata. The total number of fundamental fermions in our universe is then 1024.

The masses of the strata are individually related by the factor $\pi e^2 g_{g_{1/2}}$. We may view each fermion as composed of a multiple of its counterpart in the stratum above it. Each of the fermions in the stratum above it may be composed of a multiple of the fermions in the stratum immediately above it, and so on. Thus a stratum $n_s > 2$ fermion is composed of corresponding stratum $n_s - 1$ fermions, each of which may be composed of stratum $n_s - 2$ corresponding fermions and so on.

This view supports the Gambol Theory proposed by the author.[28] A gambol is a fraction of a fundamental fermion. Its mass is the fermion's mass divided by an integer. It has all the features of the fermion – spin, internal symmetries, … . We discuss the gambol view of strata below. Fundamental fermions are composites of gambols of other strata. The fermions of the $n_s = 2$ stratum (the ones that we know) are composed of corresponding $n_s = 1$ stratum fermions.

*Sequence Theory with its sets of fermion strata now makes gambols **Real** as manifestations of fermions of strata below the stratum of a fundamental fermion(s). Thus the 2nd stratum quarks and leptons, with which we are familiar, are made of gambols, which are real fermions of stratum 1.*

In universes of other Cosmos spaces the number of fundamental fermions is 2^{r+6} where r is the dimension. The features of Sequence Theory are the same in all universes of all spaces.

The Sequence Theory has four items (Fig. 3.4). The first row applies to the Cosmos Theory structure – setting up the form of the Cosmos Theory spaces. The remaining rows specify sequences within universes. Consequently the first row sequence coincides with the "lifetime" of Cosmos Theory structure. We have seen there

---

[28] See Blaha (2023a) for a description of Gambol Theory features.

can be no time before or after the primary r = 18 parent universe. We view the Cosmos Theory spaces as existing "forever."

The lower three items in Fig. 3.4 take effect at the creation of universes within the primary universe. Then the fundamental fermion mass and interaction structure is determined. Within a universe each mass sequence is subject to Quantum Field Theory, which modifies the masses. As we saw in previous books universes have generations and layers of fundamental fermions. The fermion masses in these structures are in sequences with the growth factors appearing in Fig. 3.4. However they will differ in their choice of the values of the constant $c_{n \circ n}$ in eq. 3.2a:

$$C_{n \circ n} = c_{n_\circ} \exp(n\pi/(\lambda - \tfrac{1}{4})^{\tfrac{1}{2}}) \qquad (3.2a)$$

Each mass sequence of each generation of each layer $c_{n \circ n}$ may have a different value. Their values are determined by detailed quantum theory calculations. Thus the masses of all fermions in all universes are determined from Sequence Theory as modified by quantum theory calculations. These calculations include Higgs boson effects.

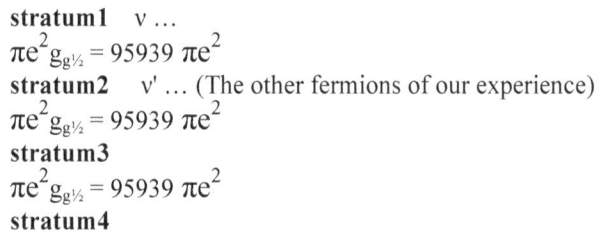

stratum1   v ...

$\pi e^2 g_{g\frac{1}{2}} = 95939 \ \pi e^2$

stratum2   v' ... (The other fermions of our experience)

$\pi e^2 g_{g\frac{1}{2}} = 95939 \ \pi e^2$

stratum3

$\pi e^2 g_{g\frac{1}{2}} = 95939 \ \pi e^2$

stratum4

Figure 4.1. A view of the four strata of the Sequence Theory fundamental fermions showing mass multipliers. Each stratum has 256 fermions in our universe (in UST theory). Taken together the 4 strata form a 2^{r+6} set of fermions in an rth dimension universe.

4.1 UST Generations and Layers of Fermions

In the Unified SuperStandard Theory there are 256 fundamental fermions arranged in four layers, each of which has four generations (different from the Standard Model). Each layer has its own set of internal symmetries and their group structures. Each layer's symmetry groups have a set of vector bosons in their fundamental representations. Each layer also has a set of scalar Higgs bosons implementing symmetry breaking.

The strata each have a set of fundamental fermions. The fermions of the strata have the same symmetry structure. They differ in mass. The strata masses form sequences based on the multiplier factor $\pi e^2 g_{g\frac{1}{2}}$. These masses are modified by quantum field theory effects such as the Higgs Mechanism. Thus their interaction dynamics are affected. Different strata have different dynamics. Yet their group symmetries are the same. *A fermion in one stratum has the same group identity and quantum numbers as the corresponding fermion in a different stratum.*

We will use these features later in our gambol discussion and in the discussion of the form of Sequence Theory quantum field theory.

4.2 Vector and Scalar Bosons

In a given universe we expect the fermions to exist as 4-fold strata. The vector boson symmetries are the same in each stratum. Their vector boson wave functions are the same for all fermion strata. A fermion in one stratum can interact with a fermion in another stratum using the ElectroWeak and Strong interactions. We discuss this later within a Gambol Theory framework.

Vector bosons are stratum independent. Vector bosons and their symmetries are not multiplied to be different in different strata. Corresponding fermions of different strata have the same interactions of the same symmetry groups. The vector bosons, being stratum independent, may undergo the same symmetry breakdown and acquire masses through the same Higgs Mechanism. The Higgs bosons are the same in all strata.[29] The masses of vector bosons are stratum independent.

4.3 Basic Algebraic Equations of Sequence Theory

Eq. 3.1 defines the basic equation of Sequence Theory. It implies the algebraic equations discussed in chapter 3. They apply to all sequences. The sequence relations that follow from eq. 3.1 are:

$$\lambda_{\text{Bohr } n\varphi} = n_\varphi^2/m \qquad\qquad (3.8a)$$
$$g_{n\varphi} = \exp(\pi/(\lambda_{\text{Bohr } n\varphi} - \tfrac{1}{4})^{\tfrac{1}{2}}) \qquad\qquad (3.8b)$$
$$\lambda_{\text{Bohr } n\varphi} = [(\ln(g_{n\varphi})/\pi]^{-2} + \tfrac{1}{4} \qquad\qquad (3.8c)$$

The sequences are characterized by

$$C_{n\varphi n} = c_{n\varphi} \exp(n\pi/(\lambda_{\text{Bohr } n\varphi} - \tfrac{1}{4})^{\tfrac{1}{2}}) \qquad\qquad (3.8d)$$

which have the initial factors $c_{n\varphi}$ for the sequences. These constants are dimensionless for sequences of dimensions. They have the dimension of [mass] for sequences of masses.

4.4 Interactions of Strata

The fermion strata all experience the same interactions of the internal symmetry groups. Corresponding interactions in the strata are the same. They are not replicated. A particle of one stratum may interact thereby with particles of other strata.

We will consider the UST set of groups for the various layers and strata. Similar considerations apply to other universes of other spaces.

In Fig. 4.2 we see the groups of each stratum are the same for the corresponding layers of each stratum. In layer 1 of each stratum the SU(4) (or SU(3)⊗U(1)) group is the same; the SU(2) ⊗U(1) group is the same; and the U(4) groups are the same. *The*

---

[29] It would be possible to have vector boson and scalar boson strata in a more general formulation. We take a minimalist approach here.

strata do not have multiples of the symmetry groups and interactions. We make this choice so that a fermion of one stratum may interact with a fermion of another stratum.

As a result we may view the fermions of one stratum as composed of the gambols (fermions) of another stratum since their symmetry group features and quantum numbers are the same. This assumption supports the author's Gambol Theory of fundamental fermions. The gambols of one stratum's fermions are the fermions of the stratum above it (excepting stratum one). See section 4.6 for more on Gambol Theory.

4.5 Strata Quantum Field Theory

In this section we will describe a Dirac equation model for the interactions of four fermion strata where each stratum has one fermion.[30] The Lagrangian is

$$\mathcal{L} = \sum_{\text{strata}} \overline{\psi}\{i\,\gamma^\mu\,\partial/\partial y^\mu + \mathcal{S}\sum_k g_k A_k^{a\mu}(y)\,\gamma_\mu T_{ka} + M\}\psi(y) + \text{c.c.} \qquad (4.2)$$

where k labels the interactions. ψ may be viewed as a strata column vector with four components (strata). The k vector boson interaction terms are the same for all four components. We implement this feature with a 4×4 matrix $\mathcal{S}$ where all elements are the integer 1. The mass matrix in this model is a 4×4 matrix for the masses of the four strata. (In this model we treat each stratum as having one fermion. In UST each stratum has 256 fermions.)

The interaction term with the $\mathcal{S}$ matrix may be rewritten as a double sum over strata:

$$\sum_{i=1}^{4}\sum_{j=1}^{4} \overline{\psi}_i \sum_k g_k A_k^{a\mu}(y)\,\gamma_\mu T_{ka}\,\psi_j + \text{c.c.} \qquad (4.3)$$

It specifies the interactions are the same for all fermions of all strata. This crucial point of the model is that it shows the interaction term applies equally to all fermions. Thus we can have an interaction with a term

$$\Psi_{n_2} \rightarrow \psi_{n_1} + \gamma \qquad (4.4)$$

where n_1 and n_2 designate strata 1 and 2 respectively, and γ represents a vector boson.

This type of model supports the concept of representing fundamental fermions as assemblages of gambols of a different stratum, which we discuss in section 4.7.

---

[30] The model generalizes directly in the case of our UST universe to 256 fermions per stratum.

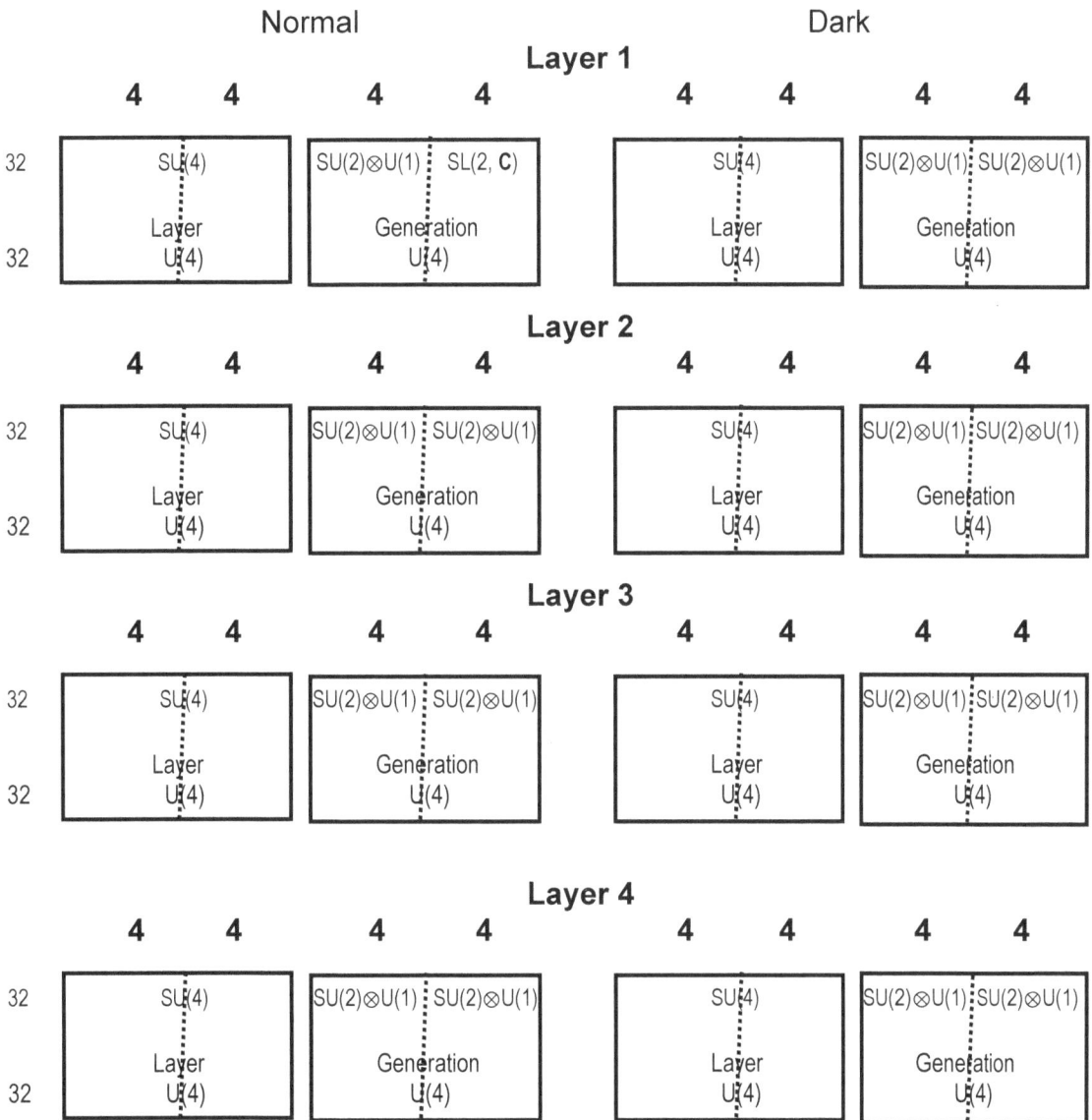

Figure 3.6. (From Blaha (2024i)). Normal and Dark symmetry groups of UST. SL(2, C) represents the Lorentz group SO+(1,3). This diagram appears in Blaha (2024i) and our earlier books such as Blaha (2020d).

Stratum 1

SU(4) U(4)	SU(2)⊗U(1¦SU(2)⊗U(1)) U¦4)
SU(4) U(4)	SU(2)⊗U(1) ¦SU(2)⊗U(1) U¦4)

Stratum 2

SU(4) U(4)	SU(2)⊗U(1¦SU(2)⊗U(1)) U¦4)
SU(4) U(4)	SU(2)⊗U(1) ¦SU(2)⊗U(1) U¦4)

Stratum 3

SU(4) U(4)	SU(2)⊗U(1¦SU(2)⊗U(1)) U¦4)
SU(4) U(4)	SU(2)⊗U(1) ¦SU(2)⊗U(1) U¦4)

Stratum 4

SU(4) U(4)	SU(2)⊗U(1¦SU(2)⊗U(1)) U¦4)
SU(4) U(4)	SU(2)⊗U(1) ¦SU(2)⊗U(1) U¦4)

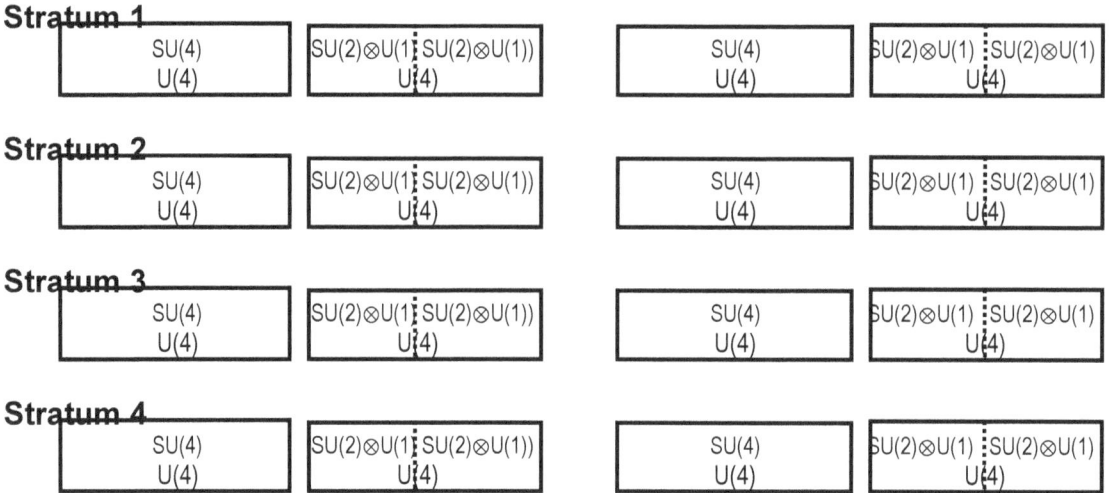

Figure 4.2. **First** layers of the four strata for fundamental fermions. Each stratum has three additional layers with similar relationships between the groups of their layers. Fig. 3.6 above shows the four layers of a single stratum. The group structures are the same for all strata.

4.6 Gambol Theory as a Result of Sequence Theory

In Blaha (2023e) and (2024a) we presented the Gambol Theory of the internal features of fundamental fermions and fermion bound states (hadrons).[31] The fermion strata of Sequence Theory provide a framework for Gambol Theory. The fermions of the first stratum that contain 256 fundamental fermions, including the neutrinos ν_e, ν_μ, and ν_τ, are without gambols. The first stratum fermions are the gambols of the second stratum. Each stratum can be viewed as composed of a multiple of gambols which are fermions of lower strata. This mapping of a fundamental fermion of one stratum to fundamental fermions of lower strata is the core feature of Gambol Theory. It is based on the equality of the symmetries and quantum numbers of corresponding fermions of all strata as verified by the form of interaction terms in eq. 4.3.

*Several years ago this author proposed Gambol Theory with the gambols being a theoretical construct – not real in the sense that quarks are real although they are only known to exist within hadrons at current energies. Sequence Theory with its sets of fermion strata now makes gambols **Real** as manifestations of fermions of strata below the stratum of a fundamental fermion(s). Thus the quarks and leptons, with which we are familiar, are made of gambols, which are real fermions of stratum 1.*

The construction of gambols within fundamental fermions takes several forms:

1. A fermion of one stratum (except stratum 1) can be mapped to a corresponding set of fermions (gambols) in the stratum below it. These

---

[31] See Blaha (2023a) and (2024a) for a description of Gambol Theory features.

gambols have identical quantum numbers with the fermion. Their mass is the mass of the fermion divided by the number of gambols within it. The basic number of gambols is the multiplier:

$$\pi e^2 g_{g_{1/2}} = 95939 \ \pi e^2 \approx 2^{15}$$

2. The number of gambols can vary by grouping gambols together in sets. Thus the number of combined gambols could be less than $95939\pi e^2$. The fundamental fermions with which we are familiar with are stratum 2 fermions. The neutrinos v_e, v_μ, and v_τ are stratum 1 fermions. Stratum 2 fermions can be viewed as an "up to 2^{15} gambol assemblage" of stratum 1 gambols. These gambols can be combined in sets down to two gambols for Gambol Theory.

3. For fermions above stratum 2 it is possible to chain together the gamboling: the fermion of a stratum may be treated as an assemblage of next lower stratum fermions (gambols), each of which can be treated as an assemblage of fermions (gambols) of the stratum two stratums below. And so on. As a consequence the number of gambols of lower strata in a fermion can be much greater than $95939 \ \pi e^2$.

4. The result is a fermion of stratum greater than 1 can consist of a large number of gambols. Each gambol has a fraction of the fermion's mass and the quantum numbers and spin of the fermion. Except for the gambol mass all gambols composing a fermion have the same properties as the fermion.

5. This procedure may be applied to hadrons composed of several quarks by creating a gambol set for each quark and then simply combining the gambols of all quarks within the hadron. See Blaha (2024a) for examples.

Example 1:

A fermion of stratum 2 (our known particles) can interact with fermions such as a neutrino of stratum 1 (or of a higher stratum) to generate an interaction like:

stratum 2 fermion → other fermions of stratum 1 plus a v_e of stratum 1   (4.5)

Example 2:

A mass M fermion of stratum 2 (our known fermions) can be viewed in Gambol Theory as composed of $\pi e^2 g_{g_{1/2}}$ gambols of mass m = $M/(\pi e^2 g_{g_{1/2}})$. We can then calculate its properties as we did in the Deep Inelastic scattering example considered in sections 4.6.1 and 4.6.2 below.

4.6.1 Deep Inelastic Lepton-Nucleon Scattering

In Blaha (2023e) we developed a model of deep inelastic lepton-nucleon scattering. The deep inelastic that was found experimentally was well approximated by our gambol formulation where the nucleon was treated as an assemblage of gambols in a Planckian distribution. It appears one can view this Gambol Theory calculation as an equivalent of a parton analysis. Below we summarize features upon which this formulation was based. See Blaha (2023e) for details.

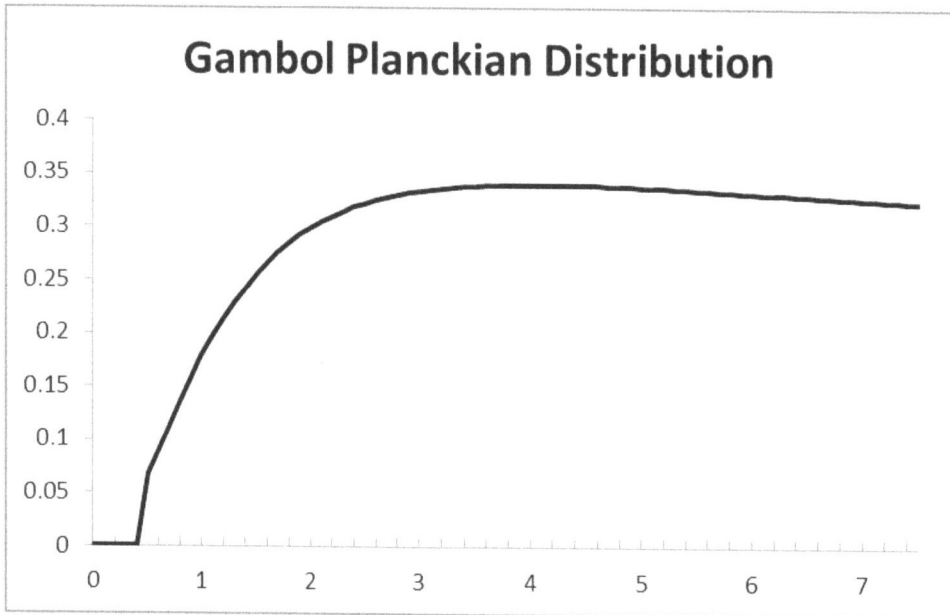

Figure 4.3. The gambol Planckian distribution plotted against ω for the R = 0 proton deep inelastic scattering.

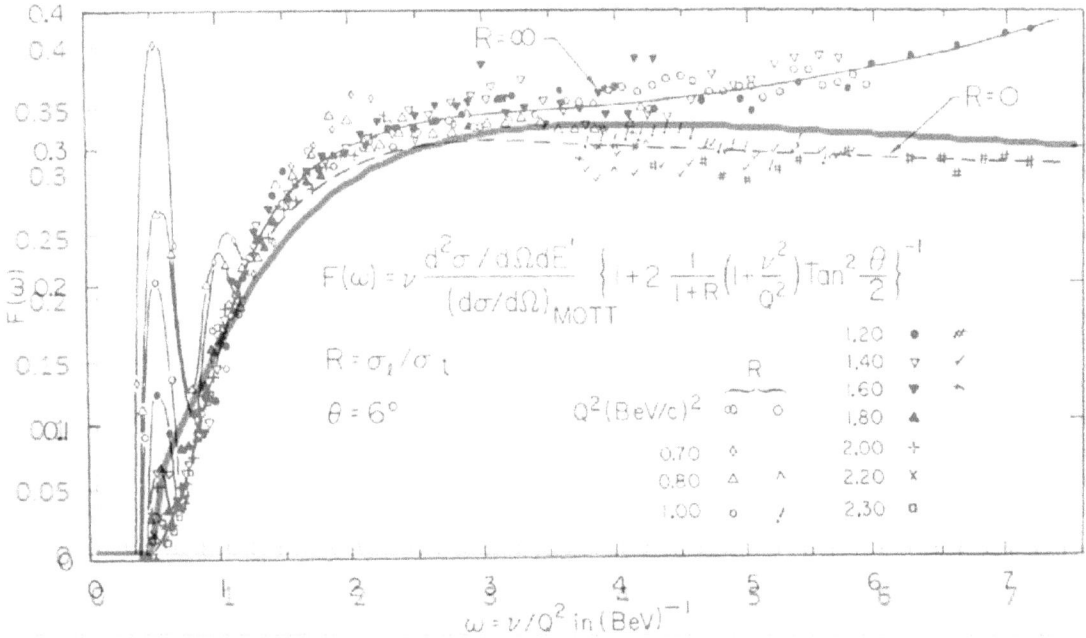

Figure 4.4. Plot of F(ω) = vW₂(ω) vs. ω = 2M/x with the gambol Planckian distribution of eq. 6.19 superimposed on it. Match!

ω	1	2	3	4	5	6	7
$F(\omega)_{experiment}$	0.19	0.3	0.3	0.3	0.3	0.3	0.3
$F(\omega)_{Planckian}$	0.18	0.299	0.33	0.34	0.34	0.33	0.326

Figure 4.5. Table of Gambol Planckian distribution values vs. experiment (noting some spread in experimental values).

4.6.2 Some Gambol Theory Principles

The Gambol Model developed in Blaha (2023e) *Cosmos Theory: The Sub-Particle Gambol Model and* 2023d, *Newton's Apple is Now the Fermion* has numerous implications which we list below. The fractionation process of a particle into sets of gambols is described in Blaha (2023d) and (2023e). Below is a Blaha (2024a) summary with some changes.

1. Fundamental Fermions, Quark and Lepton particles, are composed of probabilistically determined sets of gambols that embody the interior and dynamics of the particle.

2. The probability of each set of gambols within a fermion of mass m is determined by a Planckian distribution:

$$U_{gi}(\varepsilon(s)) = 15\, N(\pi k T_g)^{-4}\, \varepsilon^3 / (\exp(\varepsilon/(kT_g)) - 1)$$

where

$$\varepsilon(s) = [(m_g s/m + 1)/(s + 1)]E$$
$$m_g = m/s$$

is the gambol set energy $\varepsilon(s)$ with E the individual gambol dynamic energy, and m_g the gambol mass. Each set of gambols labeled s, initially numbers $s = 2^n$ gambols. Each gambol of a set constitutes a universe of the Limos sector of the Cosmos spaces. $\varepsilon(s)$ interpolates between the fermion mass and the gambol mass. $U_{gi}(\varepsilon(s))$ effectively gives the average number of gambols in a particle for each energy $\varepsilon(s)$. In the case of universes there is a modified $\varepsilon(s)$ designed to have an infinite value at the point of the Big Bang ($t = 0$).

3. The label s is made continuous in calculations.

4. Gambols are confined to the interior of fermions and bosons.

5. Gambols are fermions of a lower stratum.

6. The interior of a fermion has a gambol temperature that appears in the fermion's Planckian distribution. The temperature T_g satisfies the law

$$kT_g = 0.0785\, m$$

where m is the fermion's mass. The constant 0.0785 appears to be related to the confinement of gambols to within fermions.

In the case of universes there is a modified kT_g designed to give kT_g a time dependent value.

7. The total set of gambols in a fermion may be viewed as a gas subject to the laws of Statistical Mechanics and Thermodynamics.

8. Each boson particle acts as an individual gambol.

9. The gambols of a hadron may be viewed as an individually combined composite of the gambols of its individual quarks or as combined into an overall composite for the hadron.

10. When a fundamental fermion or hadron interacts it may be viewed as undergoing a gambol interaction with the features:

a. The interaction is a sum over sets of gambols probabilistically by individual gambols.
b. Each gambol inherits all features of the parent particle except mass and momentum which are the gambol's mass and momentum.
c. The interaction is the lowest order interaction without higher order terms for high energy interactions. We extrapolate this feature to lower energies.
d. The S-matrix for the interaction has a Planckian probability factor for each gambol set $U_g(\varepsilon(s))$:

$$S_{fi} = \sum_{Gambols} \prod_g U_g(\varepsilon(s))S_{hadg}$$

where S_{hadg} is the gambol interaction S-matrix factor.

11. Within a fundamental fermion or hadron there is an internal gambol temperature T_g. where

$$kT_g = 0.785m$$

where m is the mass of the parent particle and k is Boltzmann's constant. The 0.785 factor is shown to originate in the gambol confinement mechanism described in Blaha (2023e).

12. When a resonance is created or decays the instantaneous gambol temperature of the initiating (or end) particle and the resonance are equal, as are the energies $\varepsilon(s)$ with

$$kT_i = 0.0785\, m_{resonance}$$

13. When hadrons interact, the gambols of each particle interact with those of other particles. The resulting S-matrix is a product of Planckian gambol probabilities and an interaction S-matrix element for interacting gambols of gambol masses and momentums. These terms are summed over the individual gambols.

14. Treating universes as decaying particles of enormous mass-energy, which are composed of gambols, we define a universe gambol temperature with

$$kT_{gu} = 0.0785m_u/s$$

where s = bt gives a steadily decreasing temperature with time that mirrors the decline of the standard universe temperature to T = 2.72548° K. The constant b = 8.66685×10^{-19} sec^{-1}.

15. We take the *total* mass-energy of the universe (including Dark mass and Dark energy) to be

$$m_u = 1.712 \times 10^{81} \ \text{GeV}/c^2$$

The above points summarize the gambol models of elementary particles presented in Blaha (2023e). *They show Sequence Theory is a consistent basis of Gambol Theory.*

5. Fundamental Mass Spectrums

5.1 Fermion Mass Sequences

We decomposed the known eight fundamental fermions that appear in Fig. 3.6 into two sequences in Blaha (2024h) on p. 119:

18.2 Two Sequences of Fermion Masses

We have found two sequences of fermion masses apparent in Fig. 18.1:

	SEQUENCE 1				SEQUENCE 2		
	e	u	c	t	v'	d	s	b
Mass:	0.511×10^{-3}	1.80×10^{-3}	1.28	171	8.4×10^{-5}	4.24×10^{-3}	102×10^{-3}	4.34
Multiplier:		$2^5\pi$	$2^5\pi$	$2^5\pi$		32	32	32

where the multiplier connects adjacent masses in each sequence separately.

We suggest all four generations of fermions implement this separation into sequences. See chapter 20.

The two sequences implement the multipliers:

Up type fermion sequence multiplier: $2^5\pi = 32\pi$
Down type fermion sequence multiplier: $2^5 = 32$

These sequences have sequence multipliers that are approximations to the fermion masses. *Their role is to show the trends of the sequences.* We take the view that the deviations of the sequences' mass entries are due to perturbative effects of interactions that go beyond the basic sequences given by Sequence Theory.

Sequences C and D in section 3.8 describe the fermion sequences modified to fit within the Sequence Theory framework. The Sequence Theory multipliers, which differ from our values derived from experiment, are:

C	Fermion Sequence 2 Masses	1	$2/3 \times 2^5$	1.3
D	Fermion Sequence 1 Masses	$\frac{1}{2}^{\frac{1}{2}}$	$3/2 \times \pi 2^5$	0.65

The use of these multipliers deviates from experiments' mass values – due to quantum field theory perturbative effects.

5.2 Strata of Fermions

In any universe of any Cosmos space the fundamental fermions, which number 2^{r+4} in Cosmos Theory, are expanded by a factor of four to 2^{r+6} fermions. The corresponding fermions of each stratum have the same quantum numbers. The masses of each set of four corresponding fermions form a sequence with multiplier $\pi e^2 95939$.[32]

In our universe, the fermions of the first stratum contain 256 fundamental fermions including the neutrinos v_e, v_μ, and v_τ. The quark and lepton fermions, with which we are familiar, are in the second stratum, which has 256 fermions including the (not yet found) neutrinos v'[33] $= v_e'$, v_μ', and v_τ'. There are two additional strata – each with 256 fermions matching the UST spectrum that we have discussed in earlier work. The UST fermion spectrum expanded to four strata has the form:

STRATUM

1	v_e	v_μ	v_τ	...
2	v_e'	v_μ'	v_τ'	...
3			...	
4			...	

with each stratum having 256 fermions.

The masses of corresponding fermions in adjacent strata are in ratio $95939\pi e^2$. (subject to perturbative corrections.) The quantum numbers of corresponding fermions in the strata have the same quantum numbers.

The charged fermions in the first stratum (excepting the neutrinos) do not appear to have been observed experimentally at current energies. Thus we tentatively assume that are confined within corresponding known second stratum charged fermions. As such we treat them as gambols as described in our earlier books and in chapter 4.

The stratum 3 and 4 fermions are very massive and yet to be found. They may have gambols of lower strata within them. Their gambols may or may not be confined.

The stratum two charged fermions have gambols (and combinations of gambols) of their corresponding fermions of stratum 1. The corresponding stratum 1 (gambols) and 2 charged fermions have the same quantum numbers. The masses of the stratum 1 gambols are a fraction of the corresponding stratum 2 fermion. The baryon numbers of stratum 1 fermions (gambols) are a fraction of the corresponding stratum 2 fermion baryon's number within which it resides.

---

[32] The individual multipliers between the strata of a fermion may differ due to Quantum Field Theory perturbative effects.
[33] We proposed this neutrino in earlier books.

6. Cosmos Coupling Constant Values Related to Fundamental Fermion Masses

6.1 Sequence Theory Basis

In Blaha (2022c) we developed ProtoCosmos models for the formation of the Cosmos Theory spaces. One model was based on the assumption that a set of line segments such as we found in the Cosmos fractal curve could support the generation of the Cosmos spectrum of spaces, which was based on a power sequence of dimensions.

We chose an equation with a scaling $1/r^2$ potential since dimensions are dimensionless[34] quantities:

$$(\partial^2/\partial r^2 + 1/r \; \partial/\partial r + \partial^2/\partial\varphi^2 + \lambda/r^2 + C)\psi = 0 \qquad (3.1a)$$

Chapters 3 and 4 describe the quantum number spectrum.

6.2 The Fermion Masses Sequences

In Blaha (2024g) we found that the eight known fundamental fermions of the first layer of the four dimension Cosmos Theory space and UST could be separated into two sequences of four fermions. Sequence 2 which contained the down-type fermions was approximately estimated to have a series of ratios of $2^5 = 32$ factors between the masses. Sequence 1 fermions (up-type) had less consistent factors between masses. We tentatively estimated the ratio $2^5\pi = 32\pi$ based on an approximation to the t/c fermion mass ratio and to achieve a rough parallel with the down-type fermion factor. (Fig. 6.1)

The actual mass values of sequence 1 deviated significantly from $2^5\pi$. We attribute the deviations to additional perturbative interaction effects on masses that are beyond this model.

	SEQUENCE 1				SEQUENCE 2			
	e	u	c	t	ν'	d	s	b
Mass:	0.511×10^{-3}	1.80×10^{-3}	1.28	171	8.4×10^{-5}	4.24×10^{-3}	102×10^{-3}	4.34
Multiplier:		$2^5\pi$	$2^5\pi$	$2^5\pi$		32	32	32

Figure 6.1. The fundamental fermion mass sequences with approximate multiplier factors between the masses of each sequence.

---

[34] By dimensionless we mean that they have no dimensional unit. The Fine Structure Constant α is an example of a dimensionless constant.

6.3 The Fundamental Fermion Masses Model

The mass sequences, as pictured above, have a geometric character similar to the energy levels and the Cosmos spaces spectrum as we saw in chapter 4.

We now define a 2×2 scaling equation for *both* sequences similar to eq. 3.1a:

$$\begin{bmatrix} d^2/dr^2 + \lambda_1/r^2 - M_1 & 0 \\ 0 & d^2/dr^2 + \lambda_2/r^2 - M_2 \end{bmatrix} \begin{bmatrix} \varphi_1 \\ \varphi_2 \end{bmatrix} = 0 \qquad (6.1)$$

We related the multiplier of each sequence to a coupling constant as in chapter 4:

$$2^5\pi = \exp(\pi/(\lambda_1 - \tfrac{1}{4})^{\frac{1}{2}}) \qquad\qquad (6.2)$$
$$2^5 = \exp(\pi/(\lambda_2 - \tfrac{1}{4})^{\frac{1}{2}}) \qquad\qquad (6.3)$$

for the up-type and down-type sequences respectively. Then

$$[(\ln(32\pi)/\pi]^{-2} + \tfrac{1}{4} = \; = \lambda_1 = 0.714 \qquad \approx g_2 = 0.63 \qquad (6.4)$$

$$[(\ln(32)/\pi]^{-2} + \tfrac{1}{4} = \; = \lambda_2 = 1.07 \qquad \approx g_3 = 1.21 \qquad (6.5)$$

where we adjoin Coupling Constant g values of comparable value from the coupling constant table (Fig. 6.2).

Note the values of λ_1 and λ_2 are related by:

$$1.07/.714 = 1.5 \qquad\qquad (6.6)$$

Note the sum of the values $\lambda_1 + \lambda_2 = 1.784$ and also $g_2 + g_3 = 1.84$. These values differ by 3.4% suggesting that the ElectroWeak SU(2) and Strong SU(3) interactions combine to generate each mass sequence through an interplay of interactions.

	ESTIMATE			EXPERIMENT		
Interaction	**Expression**	**$g^2/4\pi$ Value**	**g**	**Known[35] Value $g^2/4\pi$**		**Deviation**
U(0) α_0	$e^2/4096$	0.0018036	-	-		-
U(1) $\alpha_1 = \alpha$	$e^2/1024$	0.00721438	0.303	0.0072973525643		1.15%
SU(2) $\alpha_2 = g^2/4\pi$	$e^2/256$	0.0289	0.63	0.0316		9.3%
SU(3) $\alpha_3 = \alpha_S$	$e^2/64$	0.115	1.21	0.117		1.7%
SU(4)[36] α_4	$e^2/16$	0.462	2.4?	0.458		0.087%

Figure 6.2. Coupling constants for the UST (and the Standard Model).

6.4 Separation of SU(2) and SU(3) Contributions to the Sequences' Powers

We view the values in the sequences are sums of SU(2) and SU(3) terms. We calculate the discrepancy x between the λ_i from

$$\lambda_1 + x = 1.21$$
$$\lambda_2 - x = 0.63 \qquad (6.7)$$

using the approximate equality: $\lambda_1 + \lambda_2 = 1.784$ while $g_2 + g_3 = 1.84$. We find

$$x = 0.468 \qquad (6.8)$$

.
After noting that $g_4 = 2.4$ is itself an extrapolation we will use the approximate values from Fig. 6.2

$$x = \tfrac{1}{2}(g_1 + g_2) = 0.466 \qquad (6.9)$$

Then we find an approximate relation between the SU(1), SU(2) and SU(3) coupling constants, and the λ_i, factors in in the fundamental fermion mass sequences.

$$\lambda_1 = 0.714 = 1.21 - 0.468 = g_3 - \tfrac{1}{2}(g_1 + g_2) \qquad (6.10)$$

for sequence 1, and

$$\lambda_2 = 1.07 = 0.63 + 0.468 = 3g_2/2 + \tfrac{1}{2} g_1 \qquad (6.11)$$

for sequence 2.

Sequence 2 is governed by the ElectroWeak interactions. Sequence 1 is governed by a combination of ElectroWeak and Strong SU(3) interactions. These results are reflected in the data since sequence 1 masses grow much more rapidly than sequence 2 masses.

The reasonableness of the eqs. 6.10 and 6.11 expressions for the multipliers of the sequences supports the choice of the sequence 1 multiplier $2^5\pi$ based on the t/c mass ratio, and the sequence 2 multiplier 2^5 approximately based on sequence 2 masses. They approximate the values in Fig. 3.4. The dependence of the multipliers of interactions

---

[35] All coupling constant values are based on data from Particle Data Group Tables of 2024.
[36] This value is based on the "doubling trend" seen in the three known coupling constants g above.

would appear to be directly useful for specifying the dynamics of gambols in the author's Gambol Model presented in previous books.

It is important to view the generation of fermion masses with Strong and ElectroWeak interactions as occurring *within* each individual fermion. The Higgs mechanism is used conventionally to generate vector boson masses that may play a role within interaction effects in fermions.

However the success in describing the fermion masses raises the question whether vector boson masses may also be similarly described. In addition, the masses of scalar bosons may also be so described. We have shown numerical relations between vector and scalar boson masses in Blaha (2024g) that are suggestive of this possibility. We will return to this issue in section 7.1, and in a future work.

This chapter directly relates fermion mass sequences to the coupling constant sequence.

7. The Coupling Constants Model

This chapter develops a model based on that of chapters 3 and 4 for coupling constants that generates the sequence of the four coupling constants.

We use eq. 3.1a with a scaling $1/r^2$ potential:

$$(\partial^2/\partial r^2 + 1/r\ \partial/\partial r + \partial^2/\partial\varphi^2 + \lambda/r^2 + C)\psi = 0 \qquad (3.1a)$$

where λ is a coupling constant. The eigenvalue spectrum is described in section 3.5.2. See the coupling constant table in Fig. 6.2.

The coupling constant results found in earlier books were based on counting the total number of spin terms for the fundamental fermions in a fundamental representation of the corresponding group. That approach mirrors the correspondence of the Cosmos spaces' results with counting the number of creation/annihilation operators in Fourier representations of fermion wave functions.

The coupling constant counting approach is described in Blaha (2024i):

> We have shown coupling constants appear to form a powers sequence. We now consider the expression of a coupling constant as the product of a number of factors that are based on geometry.
> First we note that n is the number of fermions in a fundamental representation of SU(n). The coupling constant $g(n) = g_n$ has the form:
>
> $$g(n) = e\ (4\pi)^{\frac{1}{2}}\ 2^{n-6} = e\ (2^{-10}\pi)^{\frac{1}{2}}\ 2^n = 0.1505\ 2^n \qquad (1.18)$$
> $$= g_g\ (\text{number of spin states per fermion})^{\text{number of fermions}}$$
> $$= g_g\ (\text{number of spin states per fermion})^n$$

The coupling constant is a fundamental value g_g times the number of spin states (a geometric property of fermions equal to half the column length of a spinor) raised to the number of fermions. The number of spin states per fermion is 2 in $r = 4$ dimensions.

> We see that g(n) has n factors – one factor for 2 spin states for each fermion in the SU(n) fundamental representation. The origin of g_g appears to be

$$g_g = e\ (2^{-10}\pi)^{\frac{1}{2}} \qquad (1.19)$$

Note

$$\alpha_n = g_n^2/4\pi = e^2\ 2^{2n-12}$$
$$= \alpha_0\ 2^{2n} \qquad (1.20)$$

Eq. 1.18 is *initially* not decisively understood due to the appearance of the natural logarithm base e. However e and π appear naturally later in our studies of higher dimension volumes and surface areas. We therefore view g_g as having a *geometric origin*.

Eq. 1.18 implies that the coupling g(n) for SU(n) is g_g times n factors of 2 which represents the number of spin states per fermion in the SU(n) fundamental representation in four dimensions. Thus it is a product of degrees of freedom.

An important point is the number of factors is equal to the number of fermions in the SU(n) fundamental representation. *The coupling constant "felt" by one fermion depends on the total number of fermions in the group's fundamental representation.*

In the present case there are four relevant internal symmetry groups and their coupling constants. The results here may be extended to higher symmetry groups (Fig. 7.1). These groups may be viewed as broken to the four symmetry groups of the UST. However we view this process as due to the structure of the Cosmos spaces – not as due to symmetry breaking.

Together with the results of chapter 6 we have a consistent view of masses, coupling constants, and the Cosmos spaces spectrum within the framework of Cosmos Theory.

	ESTIMATE			EXPERIMENT			
Interaction	Expression	$g^2/4\pi$ Value		g	Known[37] Value $g^2/4\pi$		Deviation
U(0) α_0	$e^2/4096$	0.0018036		0.15	-	-	-
U(1) $\alpha_1 = \alpha$	$e^2/1024$	0.00721438		0.303	0.0072973525643		1.15%
SU(2) $\alpha_2 = g^2/4\pi$	$e^2/256$	0.0289		0.63	0.0316		9.3%
SU(3) $\alpha_3 = \alpha_S$	$e^2/64$	0.115		1.21	0.117		1.7%
SU(4)[38] α_4	$e^2/16$	0.462		2.4?			
SU(5) α_5	$e^2/4$	1.848		4.8?			
SU(6) α_6	e^2	7.392		9.6?			
SU(7) α_7	$4e^2$	29.568		19.2?			
SU(8) α_8	$16e^2$	473.088		38.4?			
SU(9) α_9	$64e^2$	1892.352		76.8?			
SU(10) α_{10}	$256e^2$	7569.408		153.6?			

Figure 7.1. Coupling constants for the UST (and the Standard Model) extended to SU(10) by quadrupling α_i values.

---

[37] All coupling constant values are based on data from Particle Data Group Tables of 2024.
[38] The below values in the table are based on the "doubling trend" seen in the three known coupling constants g.

8. Group Interpretation of Cosmos Spaces

8.1 The Cosmos Second Order Differential Equation

The second order differential equation that has appeared repeatedly in ProtoCosmos Models and in previous chapters has a simple generic form:

$$(\partial^2/\partial r^2 + 1/r\ \partial/\partial r + \partial^2/\partial\varphi^2 + \lambda/r^2 + C)\psi = 0 \qquad (3.1a)$$

It is a second order equation in one variable (line segment) that may be viewed as part of the Cosmos Fractal Curve.[39] It has an interaction that has scaling. It has an eigenvalue that may be a mass or coupling constant or have some other role generating an eigenvalue spectrum. The scaling of the first 4 terms in eq. 3.1a leads to a power sequence set of eigenvalues.[40] The generality of this equation, and its apparent success in our studies, commends itself as fundamental.

8.2 From Where Does the Cosmos Equation Arise?

The equation's results pervade Cosmos Theory: in the specification of the Cosmos Spaces, in the Fundamental Fermion Sequences, in the Coupling Constants and apparently in the vector boson and scalar boson mass spectrums. Its universality raises the question of its origin.

The form of the differential equation eq. 3.1a shows a scaling property due to the $1/r^2$ potential. All other possible potentials have a dependence on a "dimensionful" constant.[41] The scaling property is critical in order to obtain a power spectrum sequence of C eigenvalues. This feature is the key feature of the Cosmos Theory aspects described above. It is needed in the context of the set of Cosmos Spaces to avoid a dependence of universes *of differing spaces* on the dimensions of fundamental constants.

Thus a new form of locality presents itself. *The constants determining the features of the universes of one space type are independent of the constants determining the features of the universes of a different space type. Each universe of a given space type has its own set of defining constants. Comparing details of universes of different space types then is not possible.*

---

[39] Blaha (2024b) and (2024c).
[40] Its is interesting that the Dirac equation for an atom's levels leads to an energy spectrum that is a power sequence of eigenvalues for atoms with extremely large central charge $Z\alpha \gg 1$. See K. M. Case, *op. cit.*
[41] For each sequence of powers there is an overall constant that may or may not be "dimensionful."

8.3 Power Sequence Origin of Cosmos Spaces Spectrum

We have given several ways to view the origin of the Cosmos spaces. We now explain a new view of the origin of the Cosmos spaces. The key feature of the spaces is that each space has dimension array. The dimension array size is related to the dimension r by

$$d_r = 2^{r+4}$$

In chapter 3 we determined

$$\lambda_{Bohr\ n_\varphi=4} = 16/m = \lambda_4 = 20.79 \qquad (3.10)$$

based on the power factor 2. We now note $\lambda_{Bohr\ n_\varphi=4}$ is a coupling constant value. It best matches the SU(8) coupling constant 38.4 in Fig. 7.1 since it is the lowest value greater than $\lambda_{Bohr\ n_\varphi=4}$:

$$SU(8) \qquad \alpha_8 \qquad 16e^2 \qquad 473.088 \qquad 38.4?$$

Setting

$$\lambda_{Bohr\ n_\varphi=4} + x = 38.4$$

we find

$$x = 17.61 = g_7 - 1.59 = g_7 - g_3 - g_1 - 0.08$$

where the g_i coupling constant values are from Fig. 7.1. Therefore we find the $\lambda_{Bohr\ n_\varphi=4}$ value is a mixture of SU(n) coupling constants:

$$\lambda_{Bohr\ n_\varphi=4} = g_8 - g_7 + g_3 - \tfrac{1}{2}(g_2 + g_1) + 0.08$$

The small discrepancy must be viewed as consequences of experimental uncertainties.

8.4 Structure of Cosmos Spaces

We have found a basis for Cosmos spaces in SU(8). It generates Cosmos spaces in blocks of 4×4 dimensions corresponding to its fundamental representation.

r = 2 d_2 = 64 = four 4×4 blocks
r = 4   sixteen 4×4 blocks
r = 6   sixty-four 4×4 blocks
etc.

Cosmos spaces may be viewed as composed of SU(8) fundamental representation blocks.

8.5 Composition of a Cosmos Spaces

Based on an analogy with chapter 7, which effectively treated coupling constants as "dimensionless" masses, we now treat the parts of a Cosmos space as dimensionless quantities that are composed of group parts. A group is represented within a space as a set of dimensions as we have seen before.

We saw when considering fermion masses that they were generated through ElectroWeak and Strong interactions. Here we see the dimension array of a space is generated by the groups that comprise the space through the allotment of dimensions within the dimension array for each of the groups listed for $\lambda_{\text{Bohr } n_\phi = 4}$. In a sense the groups in the list generate the dimension array. Note that there are overlaps such as the groups of g_8 and g_7 as well as the other groups. A dimension array has many possible splittings. However the SU(8) block structure does persist based on:

$$\lambda_{\text{Bohr } n_\phi = 4} = g_8 - g_7 + g_3 - \tfrac{1}{2}(g_2 + g_1)$$

The SU(7) part is broken but may be viewed in examples such as the fermions e, u, c, t, d, s, b. The neutrino is an SU(7) singlet in this view. The Strong SU(3) and ElectroWeak SU(2)⊗SU(1) parts appear in the Cosmos spaces such as the r = 4 space, and universe, of the UST and The Standard Model. These groups also appear in higher dimension spaces and universes.

REFERENCES

Akhiezer, N. I., Frink, A. H. (tr), 1962, *The Calculus of Variations* (Blaisdell Publishing, New York, 1962).

Bjorken, J. D., Drell, S. D., 1964, *Relativistic Quantum Mechanics* (McGraw-Hill, New York, 1965).

Bjorken, J. D., Drell, S. D., 1965, *Relativistic Quantum Fields* (McGraw-Hill, New York, 1965).

Blaha, S., 1995, *C++ for Professional Programming* (International Thomson Publishing, Boston, 1995).

_____, 1998, *Cosmos and Consciousness* (Pingree-Hill Publishing, Auburn, NH, 1998 and 2002).

_____, 2002, *A Finite Unified Quantum Field Theory of the Elementary Particle Standard Model and Quantum Gravity Based on New Quantum Dimensions™ & a New Paradigm in the Calculus of Variations* (Pingree-Hill Publishing, Auburn, NH, 2002).

_____, 2004, *Quantum Big Bang Cosmology: Complex Space-time General Relativity, Quantum Coordinates™ Dodecahedral Universe, Inflation, and New Spin 0, ½, 1 & 2 Tachyons & Imagyons* (Pingree-Hill Publishing, Auburn, NH, 2004).

_____, 2005a, *Quantum Theory of the Third Kind: A New Type of Divergence-free Quantum Field Theory Supporting a Unified Standard Model of Elementary Particles and Quantum Gravity based on a New Method in the Calculus of Variations* (Pingree-Hill Publishing, Auburn, NH, 2005).

_____, 2005b, *The Metatheory of Physics Theories, and the Theory of Everything as a Quantum Computer Language* (Pingree-Hill Publishing, Auburn, NH, 2005).

_____, 2005c, *The Equivalence of Elementary Particle Theories and Computer Languages: Quantum Computers, Turing Machines, Standard Model, Superstring Theory, and a Proof that Gödel's Theorem Implies Nature Must Be Quantum* (Pingree-Hill Publishing, Auburn, NH, 2005).

_____, 2006a, *The Foundation of the Forces of Nature* (Pingree-Hill Publishing, Auburn, NH, 2006).

_____, 2006b, *A Derivation of ElectroWeak Theory based on an Extension of Special Relativity; Black Hole Tachyons; & Tachyons of Any Spin.* (Pingree-Hill Publishing, Auburn, NH, 2006).

_____, 2007a, *Physics Beyond the Light Barrier: The Source of Parity Violation, Tachyons, and A Derivation of Standard Model Features* (Pingree-Hill Publishing, Auburn, NH, 2007).

_____, 2007b, *The Origin of the Standard Model: The Genesis of Four Quark and Lepton Species, Parity Violation, the ElectroWeak Sector, Color SU(3), Three Visible Generations of Fermions, and One Generation of Dark Matter with Dark Energy* (Pingree-Hill Publishing, Auburn, NH, 2007).

_____, 2008a, *A Direct Derivation of the Form of the Standard Model From GL(16) (Pingree-Hill Publishing, Auburn, NH, 2008).*

_____, 2008b, *A Complete Derivation of the Form of the Standard Model With a New Method to Generate Particle Masses Second Edition* (Pingree-Hill Publishing, Auburn, NH, 2008)

_____, 2009, *The Algebra of Thought & Reality: The Mathematical Basis for Plato's Theory of Ideas, and Reality Extended to Include A Priori Observers and Space-Time Second Edition* (Pingree-Hill Publishing, Auburn, NH, 2009).

_____, 2009a, *Bright Stars, Bright Universe* (Pingree-Hill Publishing, Auburn, NH, 2008)

_____, 2010a, *Operator Metaphysics: A New Metaphysics Based on a New Operator Logic and a New Quantum Operator Logic that Lead to a Mathematical Basis for Plato's Theory of Ideas and Reality* (Pingree-Hill Publishing, Auburn, NH, 2010).

_____, 2010b, *The Standard Model's Form Derived from Operator Logic, Superluminal Transformations and GL(16)* (Pingree-Hill Publishing, Auburn, NH, 2010).

_____, 2010c, *SuperCivilizations: Civilizations as Superorganisms* (McMann-Fisher Publishing, Auburn, NH, 2010).

_____, 2011a, *21st Century Natural Philosophy Of Ultimate Physical Reality* (McMann-Fisher Publishing, Auburn, NH, 2011).

_____, 2011b, *All the Universe! Faster Than Light Tachyon Quark Starships & Particle Accelerators with the LHC as a Prototype Starship Drive Scientific Edition* (Pingree-Hill Publishing, Auburn, NH, 2011).

_____, 2011c, *From Asynchronous Logic to The Standard Model to Superflight to the Stars* (Blaha Research, Auburn, NH, 2011).

_____, 2012a, *From Asynchronous Logic to The Standard Model to Superflight to the Stars volume 2: Superluminal CP and CPT, U(4) Complex General Relativity and The Standard Model, Complex Vierbein General Relativity, Kinetic Theory, Thermodynamics* (Blaha Research, Auburn, NH, 2012).

_____, 2012b, *Standard Model Symmetries, And Four And Sixteen Dimension Complex Relativity; The Origin Of Higgs Mass Terms* (Blaha Reasearch, Auburn, NH, 2012).

_____, 2013a, *Multi-Stage Space Guns, Micro-Pulse Nuclear Rockets, and Faster-Than-Light Quark-Gluon Ion Drive Starships* (Blaha Research, Auburn, NH, 2013).

_____, 2013b, *The Bridge to Dark Matter; A New Sibling Universe; Dark Energy; Inflatons; Quantum Big Bang; Superluminal Physics; An Extended Standard Model Based on Geometry* (Blaha Reasearch, Auburn, NH, 2013).

_____, 2014a, *Universes and Megaverses: From a New Standard Model to a Physical Megaverse; The Big Bang; Our Sibling Universe's Wormhole; Origin of the Cosmological Constant, Spatial Asymmetry of the Universe, and its Web of Galaxies; A Baryonic Field between Universes and Particles; Megaverse Extended Wheeler-DeWitt Equation* (Blaha Reasearch, Auburn, NH, 2014).

_____, 2014b, *All the Megaverse! Starships Exploring the Endless Universes of the Cosmos Using the Baryonic Force* (Blaha Research, Auburn, NH, 2014).

_____, 2014c, *All the Megaverse! II Between Megaverse Universes: Quantum Entanglement Explained by the Megaverse Coherent Baryonic Radiation Devices – PHASERs Neutron Star Megaverse Slingshot Dynamics Spiritual and UFO Events, and the Megaverse Microscopic Entry into the Megaverse* (Blaha Research, Auburn, NH, 2014).

_____, 2015a, *PHYSICS IS LOGIC PAINTED ON THE VOID: Origin of Bare Masses and The Standard Model in Logic, U(4) Origin of the Generations, Normal and Dark Baryonic Forces, Dark Matter, Dark Energy, The Big Bang, Complex General Relativity, A Megaverse of Universe Particles* (Blaha Research, Auburn, NH, 2015).

_____, 2015b, *PHYSICS IS LOGIC Part II: The Theory of Everything, The Megaverse Theory of Everything, U(4)⊗U(4) Grand Unified Theory (GUT), Inertial Mass = Gravitational Mass, Unified Extended Standard Model and a New Complex General Relativity with Higgs Particles, Generation Group Higgs Particles* (Blaha Research, Auburn, NH, 2015).

_____, 2015c, *The Origin of Higgs ("God") Particles and the Higgs Mechanism: Physics is Logic III, Beyond Higgs – A Revamped Theory With a Local Arrow of Time, The Theory of Everything Enhanced, Why Inertial Frames are Special, Universes of the Mind* (Blaha Research, Auburn, NH, 2015).

_____, 2015d, *The Origin of the Eight Coupling Constants of The Theory of Everything: U(8) Grand Unified Theory of Everything (GUTE), S^8 Coupling Constant Symmetry, Space-Time Dependent Coupling Constants, Big Bang Vacuum Coupling Constants, Physics is Logic IV* (Blaha Research, Auburn, NH, 2015).

_____, 2016a, *New Types of Dark Matter, Big Bang Equipartition, and A New U(4) Symmetry in the Theory of Everything: Equipartition Principle for Fermions, Matter is 83.33% Dark, Penetrating the Veil of the Big Bang, Explicit QFT Quark Confinement and Charmonium, Physics is Logic V* (Blaha Research, Auburn, NH, 2016).

_____, 2016b, *The Periodic Table of the 192 Quarks and Leptons in The Theory of Everything: The U(4) Layer Group, Physics is Logic VI* (Blaha Research, Auburn, NH, 2016).

_____, 2016c, *New Boson Quantum Field Theory, Dark Matter Dynamics, Dark Matter Fermion Layer Mixing, Genesis of Higgs Particles, New Layer Higgs Masses, Higgs Coupling Constants, Non-Abelian Higgs Gauge Fields, Physics is Logic VII* (Blaha Research, Auburn, NH, 2016).

_____, 2016d, *Unification of the Strong Interactions and Gravitation: Quark Confinement Linked to Modified Short-Distance Gravity; Physics is Logic VIII* (Blaha Research, Auburn, NH, 2016).

_____, 2016e, *MoND: Unification of the Strong Interactions and Gravitation II, Quark Confinement Linked to Large-Scale Gravity, Physics is Logic IX* (Blaha Research, Auburn, NH, 2016).

_____, 2016f, *CQ Mechanics: A Unification of Quantum & Classical Mechanics, Quantum/Semi-Classical Entanglement, Quantum/Classical Path Integrals, Quantum/Classical Chaos* (Blaha Research, Auburn, NH, 2016).

_____, 2016g, *GEMS Unified Gravity, ElectroMagnetic and Strong Interactions: Manifest Quark Confinement, A Solution for the Proton Spin Puzzle, Modified Gravity on the Galactic Scale* (Pingree Hill Publishing, Auburn, NH, 2016).

_____, 2016h, *Unification of the Seven Boson Interactions based on the Riemann-Christoffel Curvature Tensor* (Pingree Hill Publishing, Auburn, NH, 2016).

_____, 2017a, *Unification of the Eleven Boson Interactions based on 'Rotations of Interactions'* (Pingree Hill Publishing, Auburn, NH, 2017).

_____, 2017b, *The Origin of Fermions and Bosons, and Their Unification* (Pingree Hill Publishing, Auburn, NH, 2017).

_____, 2017c, *Megaverse: The Universe of Universes* (Pingree Hill Publishing, Auburn, NH, 2017).

_____, 2017d, *SuperSymmetry and the Unified SuperStandard Model* (Pingree Hill Publishing, Auburn, NH, 2017).

_____, 2017e, *From Qubits to the Unified SuperStandard Model with Embedded SuperStrings: A Derivation* (Pingree Hill Publishing, Auburn, NH, 2017).

_____, 2017f, *The Unified SuperStandard Model in Our Universe and the Megaverse: Quarks, ... ,* (Pingree Hill Publishing, Auburn, NH, 2017).

_____, 2018a, *The Unified SuperStandard Model and the Megaverse SECOND EDITION A Deeper Theory based on a New Particle Functional Space that Explicates Quantum Entanglement Spookiness (Volume 1)* (Pingree Hill Publishing, Auburn, NH, 2018).

_____, 2018b, *Cosmos Creation: The Unified SuperStandard Model, Volume 2, SECOND EDITION* (Pingree Hill Publishing, Auburn, NH, 2018).

_____, 2018c, *God Theory (*Pingree Hill Publishing, Auburn, NH, 2018).

_____, 2018d, *Immortal Eye: God Theory: Second Edition* (Pingree Hill Publishing, Auburn, NH, 2018).

_____, 2018e, *Unification of God Theory and Unified SuperStandard Model THIRD EDITION* (Pingree Hill Publishing, Auburn, NH, 2018).

_____, 2019a, *Calculation of: QED α = 1/137, and Other Coupling Constants of the Unified SuperStandard Theory* (Pingree Hill Publishing, Auburn, NH, 2019).

_____, 2019b, *Coupling Constants of the Unified SuperStandard Theory SECOND EDITION* (Pingree Hill Publishing, Auburn, NH, 2019).

_____, 2019c, *New Hybrid Quantum Big_Bang–Megaverse_Driven Universe with a Finite Big Bang and an Increasing Hubble Constant* (Pingree Hill Publishing, Auburn, NH, 2019).

_____, 2019d, *The Universe, The Electron and The Vacuum* (Pingree Hill Publishing, Auburn, NH, 2019).

_____, 2019e, *Quantum Big Bang – Quantum Vacuum Universes (Particles)* (Pingree Hill Publishing, Auburn, NH, 2019).

_____, 2019f, *The Exact QED Calculation of the Fine Structure Constant Implies ALL 4D Universes have the Same Physics/Life Prospects* (Pingree Hill Publishing, Auburn, NH, 2019).

_____, 2019g, *Unified SuperStandard Theory and the SuperUniverse Model: The Foundation of Science* (Pingree Hill Publishing, Auburn, NH, 2019).

_____, 2020a, *Quaternion Unified SuperStandard Theory (The QUeST) and Megaverse Octonion SuperStandard Theory (MOST)* (Pingree Hill Publishing, Auburn, NH, 2020).

_____, 2020b, *United Universes Quaternion Universe - Octonion Megaverse* (Pingree Hill Publishing, Auburn, NH, 2020).

_____, 2020c, *Unified SuperStandard Theories for Quaternion Universes & The Octonion Megaverse* (Pingree Hill Publishing, Auburn, NH, 2020).

_____, 2020d, *The Essence of Eternity: Quaternion & Octonion SuperStandard Theories* (Pingree Hill Publishing, Auburn, NH, 2020).

_____, 2020e, *The Essence of Eternity II* (Pingree Hill Publishing, Auburn, NH, 2020).

_____, 2020f, *A Very Conscious Universe* (Pingree Hill Publishing, Auburn, NH, 2020).

_____, 2020g, *Hypercomplex Universe* (Pingree Hill Publishing, Auburn, NH, 2020).

_____, 2020h, *Beneath the Quaternion Universe* (Pingree Hill Publishing, Auburn, NH, 2020).

_____, 2020i, *Why is the Universe Real? From Quaternion & Octonion to Real Coordinates* (Pingree Hill Publishing, Auburn, NH, 2020).

_____, 2020j, *The Origin of Universes: of Quaternion Unified SuperStandard Theory (QUeST); and of the Octonion Megaverse (UTMOST)* (Pingree Hill Publishing, Auburn, NH, 2020).

_____, 2020k, *The Seven Spaces of Creation: Octonion Cosmology* (Pingree Hill Publishing, Auburn, NH, 2020).

_____, 2020l, *From Octonion Cosmology to the Unified SuperStandard Theory of Particles* (Pingree Hill Publishing, Auburn, NH, 2020).

_____, 2021a, *Pioneering the Cosmos* (Pingree Hill Publishing, Auburn, NH, 2021).

_____, 2021b, *Pioneering the Cosmos II* (Pingree Hill Publishing, Auburn, NH, 2021).

_____, 2021c, *Beyond Octonion Cosmology* (Pingree Hill Publishing, Auburn, NH, 2021).

_____, 2021d, *Universes are Particles* (Pingree Hill Publishing, Auburn, NH, 2021).

_____, 2021e, *Octonion-like dna-based life, Universe expansion is decay, Emerging New Physics* (Pingree Hill Publishing, Auburn, NH, 2021).

_____, 2021f, *The Science of Creation New Quantum Field Theory of Spaces* (Pingree Hill Publishing, Auburn, NH, 2021).

_____, 2021g, *Quantum Space Theory With Application to Octonion Cosmology & Possibly To Fermionic Condensed Matter* (Pingree Hill Publishing, Auburn, NH, 2021).

_____, 2021h, *21st Century Natural Philosophy of Octonion Cosmology , and Predestination, Fate, and Free Will* (Pingree Hill Publishing, Auburn, NH, 2021).

_____, 2021i, *Beyond Octonion Cosmology II : Origin of the Quantum; A New Generalized Field Theory (GiFT); A Proof of the Spectrum of Universes; Atoms in Higher Universes* (Pingree Hill Publishing, Auburn, NH, 2021).

_____, 2021j, *Integration of General Relativity and Quantum Theory: Octonion Cosmology, GiFT, Creation/Annihilation Spaces CASe, Reduction of Spaces to a Few Fermions and Symmetries in Fundamental Frames* (Pingree Hill Publishing, Auburn, NH, 2021).

_____, 2022a, *New View of Octonion Cosmology Based on the Unification of General Relativity and Quantum Theory* (Pingree Hill Publishing, Auburn, NH, 2022).

_____, 2022b, *The Dust Beneath Hypercomplex Cosmology* (Pingree Hill Publishing, Auburn, NH, 2022).

_____, 2022c, *Passing Through Nature to Eternity: ProtoCosmos, HyperCosmos, Unified SuperStandard Theory* (Pingree Hill Publishing, Auburn, NH, 2022).

_____, 2022d, *HyperCosmos Fractionation and Fundamental Reference Frame Based Unification: Particle Inner Space Basis of Parton and Dual Resonance Models* (Pingree Hill Publishing, Auburn, NH, 2022).

_____, 2022e, *A New UniDimension ProtoCosmos and SuperString F-Theory Relation to the HyperCosmos* (Pingree Hill Publishing, Auburn, NH, 2022).

_____, 2022f, *The Cosmic Panorama: ProtoCosmos, HyperCosmos,Unified SuperStandard Theory (UST) Derivation* (Pingree Hill Publishing, Auburn, NH, 2022).

_____, 2022g, *Ultimate Origin: ProtoCosmos and HyperCosmos* (Pingree Hill Publishing, Auburn, NH, 2022).

_____, 2023a, *UltraUnification and the Generation of the Cosmos* (Pingree Hill Publishing, Auburn, NH, 2023).

_____, 2023b, *God and and Cosmos Theory* (Pingree Hill Publishing, Auburn, NH, 2023).

_____, 2023c, *A New Completely Geometric SU(8) Cosmos Theory; New PseudoFermion Fields; Fibonacci-like Dimension Arrays; Ramsey Number Approximation* (Pingree Hill Publishing, Auburn, NH, 2023).

_____, 2023d, *Newton's Apple is Now the Fermion* (Pingree Hill Publishing, Auburn, NH, 2023).

_____, 2023e,*Cosmos Theory: The Sub-Particle Gambol Model* (Pingree Hill Publishing, Auburn, NH, 2023).

_____, 2024a, *Cosmos-Universe-Particle-Gambol Theory* (Pingree Hill Publishing, Auburn, NH, 2024).

_____, 2024b, *Fractal Cosmos Theory* (Pingree Hill Publishing, Auburn, NH, 2024).

_____, 2024c, *Fractal Cosmic Curve: Tensor-Based CosmosTheory* (Pingree Hill Publishing, Auburn, NH, 2024).

_____, 2024d, *The Eternal Form of Cosmos Theory* (Pingree Hill Publishing, Auburn, NH, 2024).

_____, 2024e, *The Eternal Form of Cosmos Theory Third Edition* (Pingree Hill Publishing, Auburn, NH, 2024).

_____, 2024f, *Fundamental Constants of Cosmos Theory and The Standard Model* (Pingree Hill Publishing, Auburn, NH, 2024).

_____, 2024g, *Quark, Lepton, W and Z Masses of Cosmos Theory and The Standard Model* (Pingree Hill Publishing, Auburn, NH, 2024).

_____, 2024h, *Geometric Cosmos Geometric Universe* (Pingree Hill Publishing, Auburn, NH, 2024).

_____, 2024i, *Particles and Universes of Cosmos Theory* (Pingree Hill Publishing, Auburn, NH, 2024).

_____, 2024j, *Unification of the Subluminal and the Superluminal in Cosmos Theory* (Pingree Hill Publishing, Auburn, NH, 2024).

_____, 2024k, *The Dawn of Dynamic Cosmos Dimension Arrays* (Pingree Hill Publishing, Auburn, NH, 2024).

_____, 2024l, *Structure and Dynamics of Cosmos Theory and the Unified SuperStandard Theory* (Pingree Hill Publishing, Auburn, NH, 2024).

_____, 2025a, *Black Holes, White Holes, and Superluminal Starship* (Pingree Hill Publishing, Auburn, NH, 2025).

Eddington, A. S., 1952, *The Mathematical Theory of Relativity* (Cambridge University Press, Cambridge, U.K., 1952).

Fant, Karl M., 2005, *Logically Determined Design: Clockless System Design With NULL Convention Logic* (John Wiley and Sons, Hoboken, NJ, 2005).

Feinberg, G. and Shapiro, R., 1980, *Life Beyond Earth: The Intelligent Earthlings Guide to Life in the Universe* (William Morrow and Company, New York, 1980).

Gelfand, I. M., Fomin, S. V., Silverman, R. A. (tr), 2000, *Calculus of Variations* (Dover Publications, Mineola, NY, 2000).

Giaquinta, M., Modica, G., Souchek, J., 1998, *Cartesian Coordinates in the Calculus of Variations* Volumes I and II (Springer-Verlag, New York, 1998).

Giaquinta, M., Hildebrandt, S., 1996, *Calculus of Variations* Volumes I and II (Springer-Verlag, New York, 1996).

Gradshteyn, I. S. and Ryzhik, I. M., 1965, *Table of Integrals, Series, and Products* (Academic Press, New York, 1965).

Heitler, W., 1954, *The Quantum Theory of Radiation* (Claendon Press, Oxford, UK, 1954).

Huang, Kerson, 1992, *Quarks, Leptons & Gauge Fields 2nd Edition* (World Scientific Publishing Company, Singapore, 1992).

Jost, J., Li-Jost, X., 1998, *Calculus of Variations* (Cambridge University Press, New York, 1998).

Kaku, Michio, 1993, *Quantum Field Theory*, (Oxford University Press, New York, 1993).

Kirk, G. S. and Raven, J. E., 1962, *The Presocratic Philosophers* (Cambridge University Press, New York, 1962).

Landau, L. D. and Lifshitz, E. M., 1987, *Fluid Mechanics 2nd Edition*, (Pergamon Press, Elmsford, NY, 1987).

Rescher, N., 1967, *The Philosophy of Leibniz* (Prentice-Hall, Englewood Cliffs, NJ, 1967).

Riesz, Frigyes and Sz.-Nagy, Béla, 1990, *Functional Analysis* (Dover Publications, New York, 1990).

Sakurai, J. J., 1964, *Invariance Principles and Elementary Particles* (Princeton University Press, Princeton, NJ, 1964).

Weinberg, S., 1972, *Gravitation and Cosmology* (John Wiley and Sons, New York, 1972).

Weinberg, S., 1995, *The Quantum Theory of Fields Volume I* (Cambridge University Press, New York, 1995).

INDEX

About the Author

Stephen Blaha is a well-known Physicist and Man of Letters with interests in Science, Society and civilization, the Arts, and Technology. He had an Alfred P. Sloan Foundation scholarship in college. He received his Ph.D. in Physics from Rockefeller University. He has served on the faculties of several major universities. He was also a Member of the Technical Staff at Bell Laboratories, a manager at the Boston Globe Newspaper, a Director at Wang Laboratories, and President of Blaha Software Inc. and of Janus Associates Inc. (NH).

Among other achievements he was a co-discoverer of the "r potential" for heavy quark binding developing the first (and still the only demonstrable) non-Aeolian gauge theory with an "r" potential; first suggested the existence of topological structures in superfluid He-3; first proposed Yang-Mills theories would appear in condensed matter phenomena with non-scalar order parameters; first developed a grammar-based formalism for quantum computers and applied it to elementary particle theories; first developed a new form of quantum field theory without divergences (thus solving a major 60 year old problem that enabled a unified theory of the Standard Model and Quantum Gravity without divergences to be developed); first developed a formulation of complex General Relativity based on analytic continuation from real space-time; first developed a generalized non-homogeneous Robertson-Walker metric that enabled a quantum theory of the Big Bang to be developed without singularities at t = 0; first generalized Cauchy's theorem and Gauss' theorem to complex, curved multidimensional spaces; received Honorable Mention in the Gravity Research Foundation Essay Competition in 1978; first developed a physically acceptable theory of faster-than-light particles; first derived a composition of extremums method in the Calculus of Variations; first quantitatively suggested that inflationary periods in the history of the universe were not needed; first proved Gödel's Theorem implies Nature must be quantum; provided a new alternative to the Higgs Mechanism, and Higgs particles, to generate masses; first showed how to resolve logical paradoxes including Gödel's Undecidability Theorem by developing Operator Logic and Quantum Operator Logic; first developed a quantitative harmonic oscillator-like model of the life cycle, and interactions, of civilizations; first showed how equations describing superorganisms also apply to civilizations. A recent book shows his theory applies successfully to the past 14 years of history and to *new* archaeological data on Andean and Mayan civilizations as well as Early Anatolian and Egyptian civilizations.

He first developed an axiomatic derivation of the form of The Standard Model from geometry – space-time properties – The Unified SuperStandard Model. It unifies all the known forces of Nature. It also has a Dark Matter sector that includes a Dark ElectroWeak sector with Dark doublets and Dark gauge interactions. It uses quantum coordinates to remove infinities that crop up in most interacting quantum field theories and additionally to remove the infinities that appear in the Big Bang and generate inflationary growth of the universe. It shows gravity has a MOND-like form without sacrificing Newton's Laws. It relates the interactions of the MOND-like sector of gravity with the r-potential of Quark Confinement. The axioms of the theory lead to the question of their origin. We suggest in the preceding edition of this book it can be attributed to an entity with God-like properties. We explore these properties in "God Theory" and show they predict that the Cosmos exists forever although individual universes (or incarnations of our universe) "come and go." Several other important results emerge from God Theory such a functionally triune God. The Unified SuperStandard Theory has many other important parts described in the Current Edition of *The Unified SuperStandard Theory* and expanded in subsequent volumes.

Blaha has had a major impact on a succession of elementary particle theories: his Ph.D. thesis (1970), and papers, showed that quantum field theory calculations to all orders in ladder approximations could not give scaling deep inelastic electron-nucleon scattering. He later showed the eigenvalue equation for the fine structure constant α in Johnson-Baker-Willey QED had a zero at $\alpha = 1$ not 1/137 by solving the Schwinger-Dyson equations to all orders in an approximation that agreed with exact results to 4th order in α thus ending interest in this theory. In 1979 at Prof. Ken Johnson's (MIT) suggestion he calculated the proton-neutron mass difference in the MIT bag model and found the result had the wrong sign reducing interest in the bag model. These results all appear in Physical Review papers. In the 2000's he repeatedly pointed out the shortcomings of SuperString theory and showed that The Standard Model's form could be derived from space-time geometry by an extension of Lorentz transformations to faster than light transformations. This deeper space-time basis greatly increases the possibility that it is part of THE fundamental theory. Recently, Blaha showed that the Weak interactions differed significantly from the Strong, electromagnetic and gravitation interactions in important respects while these interactions had similar features, and suggested that ElectroWeak theory, which is essentially a glued union of the Weak interactions and Electromagnetism, possibly modulo unknown Higgs particle features, be replaced by a unified theory of the other interactions combined with a stand-alone Weak interaction theory. Blaha also showed that, if Charmonium calculations are taken seriously, the Strong interaction coupling constant is only a factor of five larger than the electromagnetic coupling constant, and thus Strong interaction perturbation theory would make sense and yield physically meaningful results.

In graduate school (1965-71) he wrote substantial papers in elementary particles and group theory: The Inelastic E- P Structure Functions in a Gluon Model. Phys. Lett. B40:501-502,1972; Deep-Inelastic E-P Structure Functions In A Ladder Model With Spin 1/2 Nucleons, Phys.Rev. D3:510-523,1971; Continuum Contributions To The Pion Radius, Phys. Rev. 178:2167-2169,1969; Character Analysis of U(N) and SU(N), J. Math. Phys. <u>10</u>, 2156 (1969); and The Calculation of the Irreducible Characters of the Symmetric Group in Terms of the Compound Characters, (Published as Blaha's Lemma in D. E. Knuth's book: *The Art of Computer Programming Vols. 1 – 4*).

In the early 1980's Blaha was also a pioneer in the development of UNIX for financial, scientific and Internet applications: benchmarked UNIX versions showing that block size was critical for UNIX performance, developing financial modeling software, starting database benchmarking comparison studies, developing Internet-like UNIX networking (1982) and developing a hybrid shell programming technique (1982) that was a precursor to the PERL programming language. He was also the manager of the AT&T ten-year future products development database. His work helped lead to commercial UNIX on computers such as Sun Micros, IBM AIX minis, and Apple computers.

In the 1980's he pioneered the development of PC Desktop Publishing on laser printers and was nominated for three "Awards for Technical Excellence" in 1987 by PC Magazine for PC software products that he designed and developed.

Recently he has developed a theory of Megaverses – actual universes of which our universe is one – with quantum particle-like properties based on the Wheeler-DeWitt equation of Quantum Gravity. He has developed a theory of a baryonic force, which had been conjectured many years ago, and estimated the strength of the force based on discrepancies in measurements of the gravitational constant G. This force, operative in D-dimensional space, can be used to escape from our universe in "uniships" which are the equivalent of the faster-than-light starships proposed in the author's earlier books. Thus travel to other universes, as well as to other stars is possible.

Blaha also considered the complexified Wheeler-DeWitt equation and showed that its limitation to real-valued coordinates and metrics generated a Cosmological Constant in the Einstein equations.

The author has also recently written a series of books on the serious problems of the United States and their solution as well as a book on the decline of Mankind that will follow from current social and genetic trends in Mankind.

In the past twenty years Dr. Blaha has written over 80 books on a wide range of topics. Some recent major works are: *From Asynchronous Logic to The Standard Model to Superflight to the Stars, All the Universe!, SuperCivilizations: Civilizations as Superorganisms, America's Future: an Islamic Surge, ISIS, al Qaeda, World Epidemics, Ukraine, Russia-China Pact, US Leadership Crisis, The Rises and Falls of Man – Destiny – 3000 AD: New Support for a Superorganism MACRO-THEORY of CIVILIZATIONS From CURRENT WORLD TRENDS and NEW Peruvian, Pre-Mayan, Mayan, Anatolian, and Early Egyptian Data, with a Projection to 3000 AD,* and *Mankind in Decline: Genetic Disasters, Human-Animal Hybrids, Overpopulation, Pollution, Global Warming, Food and Water Shortages, Desertification, Poverty, Rising Violence, Genocide, Epidemics, Wars, Leadership Failure.*

He has taught approximately 4,000 students in undergraduate, graduate, and postgraduate corporate education courses primarily in major universities, and large companies and government agencies.

He developed a quantum theory, The Unified SuperStandard Theory (UST), which describes elementary particles in detail without the difficulties of conventional quantum field theory. He found that the internal symmetries of this theory could be exactly derived from an octonion theory called QUeST. He further found that another octonion theory (UTMOST) describes the Megaverse. It can hold QUeST universes such as our own universe. It has an internal symmetry structure which is a superset of the QUeST internal symmetries.

Recently he developed Octonion Cosmology. He replaced it with HyperCosmos theory, which has significantly better features. He developed a fractionalization process for dimensions, particles and symmetry groups. He also described transformation that reduced particles and dimensions to a far more compact form. He also developed a precursor theory ProtoCosmos that leads to the HyperCosmos.

The author showed that space-time and Internal Symmetries can be unified in any of the ten HyperCosmos spaces in their associated HyperUnification spaces. The combined set of HyperUnification spaces enable all HyperCosmos dimensions to be obtained by a General Relativistic transformation from one primordial dimension in the 42 space-time dimension unified HyperUnification space.

At present the author devel;oped the Cosmos Theory that incorporates ProtoCosmos Theory, HyperCosmos Theory, Limos Theory, Second Kind HyperCosmos Theory and HyperUnification Spaces. He has introduced PseudoFermion wave functions and theory, He has related Cosmos Theory to Regge trajectories of spaces, parton theory, Veneziano amplitudes, Fibonacci numbers and Ramsey numbers. He has calculated an approximation to the difficult R(n,n) Ramsey numbers.

He has developed a Gambol Model that successfully accounts for e-p deep inelastic scattering, fundamental particle resonances, hadron scattering, and the inner structure of particles based on confinement through Casimir forces of ideal gambol gases. The Gambol Planckian Distribution was derived.

He has applied the Gambol Model to particles, universes, and the Cosmos of universes. He showed that the Cosmos may have a distribution of 23 universes corresponding to various Cosmos spaces.

Recently he showed that Cosmos Theory follows from the number of independent asymmetric tensors in a dimension r. He also showed the close parallel between the form of γ-matrices and Cosmos Theory dimension arrays. The closeness suggested that dimension arrays have the same importance as γ-matrices for fermions.

He demonstrated that the pressure of fermions within a space of dimension r balances the Casimir vacuum energy force for 18 dimensions. He showed that $2e\pi = 17.02$ marks the critical point where pressure balances Casimir force, which implies $r = 18$ is the highest dimension Physical Cosmos space. The dimension $2e\pi$ appears to set the approximate dimension for Cosmos spaces with dimension array size $2^{r+4} \cong (17.02/8)^{r+4} \cong (e\pi/4)^{r+4} \cong 2.13^{r+4}$.

Now he has found the sequences of Coupling Constant values and fermion masses in the Standard Model and UST. This book unifies the Cosmos spaces spectrum, the coupling constant spectrum and the fermion mass sequences in a fundamental scaling differential equation.

www.ingramcontent.com/pod-product-compliance
Lightning Source LLC
Chambersburg PA
CBHW040138200326
41458CB00025B/6301